HOW TO WIN A GRAND PRIX

HOW TO WIN A GRAND PRIX

FROM PIT LANE TO PODIUM – THE INSIDE TRACK

BERNIE COLLINS

Written with Maurice Hamilton

QUERCUS

First published in Great Britain in 2024 by

QUERCUS

Quercus Editions Ltd
Carmelite House
50 Victoria Embankment
London EC4Y 0DZ
An Hachette UK company

A CIP catalogue record for this book is available
from the British Library

HB ISBN 978 1 52943 759 1
TPB ISBN 978 1 52943 760 7
EBOOK ISBN 978 1 52943 762 1

Picture credits (in order of appearance): 1, 2 – Queen's University Belfast;
3, 4, 5 – McLaren Racing; 6 – McLaren GT; 7, 8, 13, 17 – Author's own;
9, 10, 12, 15 – James Moy; 11 –RaceWatch race control software by SBG Software,
a Catapult company; 14, 18 – Aston Martin Racing; 16 – Conor McDonnell;
19, 20 – Charts generated by the author.
Charts designed by Nick May

Typeset in Rotis Serif by CC Book Production
Printed and bound in Great Britain by Clays Ltd, Elcograf S.p.A.

Papers used by Quercus are from well-managed forests and other responsible sources.

Dedicated to Nanny Nell.

Contents

Chapter 1

FERMANAGH TO FORMULA 1

When growing up in County Fermanagh, I never imagined I would end up spending my time travelling the world and working as a strategist in Formula 1, never mind being part of a winning grand prix team.

There is a lot of good motorsport in Northern Ireland – rallying, motorbikes, circuit racing and so on – but my family was not particularly involved. I do remember my dad watching F1 races on television. I was aware of the drivers, the cars and the pit crews but, beyond that, I had no idea about the role being played by designers and engineers, and the interplay between them and a team's strategist.

It never occurred to me that I might one day enter that world, spend 136 races on the pit wall for Aston Martin (also known, in former guises, as Force India and Racing Point), having worked 22 races before that as a performance engineer with McLaren Racing, bringing a total of nine podium finishes across nine seasons.

The pinnacle – in every sense – was Sergio 'Checo' Pérez reaching the top step of the podium for Racing Point-Mercedes in the 2020 Sakhir Grand Prix in Bahrain. It seems incredible now, inconceivable to a younger Bernie living in a quiet rural setting just outside the town of Maguiresbridge.

My first school, in the village of Lisbellaw, was about as far removed from the hectic world of motorsport as you could ever imagine. There were around 50 kids in the entire primary school. Known as Tattygar, there were only two teachers. It eventually expanded to three!

It was a bit of a shock when I eventually moved on to Mount Lourdes, an all-girls grammar school in Enniskillen, the county town of Fermanagh. That was a totally different experience. There were 125 pupils in my year; 1,000 in the entire school. I was 11 years old and a bit lost. I just kept doing subjects that I enjoyed; I didn't really have a plan. At that stage, I didn't have a lot of exposure to engineering; I didn't really understand an engineer's role. And yet, when I look back now, there were signs that I was destined to do something in that sphere because I was forever taking things apart. I was really interested in how things worked.

But when it came to deciding my future, I had no clear idea. It was only when going through the prospectus for Queen's University in Belfast that I was drawn to mechanical engineering, mainly because it was a broad subject with some interesting possibilities – mechanical engineers can be involved in anything: all kinds of work and research, from motorsport, to outer space to hip replacements. A turning point came when I joined the university's Formula Student team in 2007.

Run by the Institution of Mechanical Engineers (IMechE), Formula Student is part of a degree-level project, the idea being to provide a standard for engineering students to meet before transitioning from university to the workplace. It is similar to F1 insofar as universities compete against each other by designing, building and racing little single-seater race cars. A set of regulations must be followed in the same way F1 teams are restricted by detailed rules. But the Formula Student regulations have enough scope to allow universities to come up with differing and innovative solutions.

Studying a Masters at Queen's University meant four years of academics: in the third year, students completed an individual project; in your final year, it was necessary to do a group project. Queen's allowed Formula Student to be used in both cases to go towards your degree, which was good because not all universities did that.

Not only did your team need to design and build a car from scratch, but funding also had to be raised from whatever sources you could find, be that local sponsorship or, perhaps, money from the university. You worked as a team of seven or eight using Formula Student as part of everyone's group project, with perhaps as many as ten third-year student individual projects going into it.

For my individual project, I designed the brakes. Individual designs did not necessarily need to make it to the car; they were more like a research project. But if your work was good, it could go on the car. The brake discs were usually cross-drilled, but I decided to go for a slotted version which, on paper, would be lighter and therefore better. These were fitted to the car. The next stage would be competing on the track.

The British leg of Formula Student was against other universities at Silverstone. There were four events: straight line; figure of eight; a sprint involving a quick lap; and an endurance event of around 25 laps. Each team was marked on design, cost, marketing and, of course, results in the four events. In many ways, these were the same pieces of the puzzle any motorsport team – from junior formulae right up to F1 – had to deal with.

Before taking part in any of the events, it was necessary to pass scrutineering to prove the car was safe and conformed with the regulations. Part of the scrutineering process involved a test in which all four wheels had to be locked up under hard braking.

My brake disc broke. This not only destroyed the brake but also a wheel rim and part of the suspension. As the junior member of the team I felt very guilty, because my individual project had gone on the car and then destroyed some of the senior members' work. A lot of hard work had gone into the car – and it had got no further than scrutineering.

In what I would later discover to be typical motor racing fashion, you simply got your head down and found a solution. It meant working late nights because we did not have many spare parts and it would be necessary to revert to standard brake discs. We had to search locally for supplies and bodge something together. We got through scrutineering, but we didn't do well in the events. Saying that, we got there in the end.

We weren't alone in having difficulty with the brake test. Locking up all four wheels is not something that happens under normal circumstances. But that was the test. The team next to us at Silverstone was flashy compared to our modest outfit. I remember, among other things, they had Red Bull cans with their names on them. We were scraping by with sandwiches and drinks from Tesco.

This neighbouring team was very advanced. They had designed an ABS system, which was all very well but it meant there was no way they could lock up their brakes! ABS is designed to prevent exactly that. They failed the test for a completely different reason. But there had to be a lesson in there: probably something to do with reading the regulations thoroughly in advance – a golden rule in any competition.

There was another team nearby who had arrived in their own truck complete with a lathe in the back. By comparison, we had hired a van and driven it there! But everyone was very helpful; they let us use the lathe and it added to the fantastic experience of making your own car and being pitted against some serious competition.

All told, Formula Student provided a really useful and revealing mix of engineering experience and insight into how to run a mini-company. You needed to work as a team to make sure the chassis and components were ready in time; everything needed to come together while working to a very low budget. Then you arrive at Silverstone and see all the other cars; some are in better shape than yours, and some are worse. It's an interesting scenario.

Twelve months later, we were back. I was team leader for my final year as part of the group project. Previously, we had made a space-frame chassis but, for the first time, we decided to do a carbon-fibre chassis, mainly because one of the team members, Findhan, was very into his carbon fibre. So we went along with that, each team member concentrating on a particular aspect. This time, I was doing suspension.

I found the team leader position to be straightforward because everyone worked really well together. In my final year I probably spent 80 per cent of my time doing Formula Student rather than the other classes, even though they were contributing significantly more to my degree. But Formula Student continued to be a really useful experience. When we went back to Silverstone, we brought both the new car and the one from the previous year.

I was also one of the drivers; in fact, of the eight team members, everyone had a drive at some stage across the four events. Because I was the lightest team member, it made sense for me to do the acceleration event. There was a regulation covering the position of the steering rack in association with the pedals. But because I was quite short, I was to discover – quite late as it would turn out – that the pedals were almost out of reach.

As often happens in many things these days, we had an electronics failure. It tends to be complicated, added to which there is usually only one person who truly understands what the electrical wiring loom is doing. And then you find that the loom is one of the last parts to be fitted because there always seems to be a wait for other stuff to be done.

I had used the older car for practice. This car had quick shifts; effectively, using full throttle while up-shifting through the gears. There was a restriction on engines; most teams chose the Yamaha R6. This was a motorbike engine, the gearshift was sequential and you didn't have to use the clutch. It was not massive in terms of speed or power output, but it was just right for what we needed. There was a rev counter, and the lights showed just when to shift. Easy!

The new car had all of this, plus launch control (to help with a fast getaway and prevent wheelspin off the line). When I finally got in, none of the tech I had practised with was working. I was sitting on the start/finish straight and one of the guys leaned in and said, 'It's fine. Just shift when you hear the rev-limiter.' When I told him I'd never hit the rev limiter in a car in my entire life, he said, 'You'll be okay. You'll know when it happens!' Really? Thanks!

I remember taking off, hearing a '*BRRRR*' and thinking, 'Ah! Okay, got to shift now!' That was fine until I got to the end of the straight – and discovered the pedals were too far away for me to brake heavily. Even so, we had managed to get going – and to stop. Just about! This was not the way to do it. It was an object lesson in getting stuff ready in time to test it.

We did two runs, but I only remember the first! I am sure the second run was better, because I knew what I was doing. All of this was on the old start/finish straight, two weeks after the British Grand Prix. Formula Student had brought me to Silverstone for the first time. Little did I realise the significant part this place was to play in my life.

Formula Student was a really good project which allowed me to see a potential career. I could do the academic stuff and I could do the design work. I could figure out the best way to make something work. There were also some practical and on-track aspects that I had not considered before because, as a mechanical engineer in Northern Ireland, there were a lot of aerospace and ship-building opportunities, but not many people were going into motorsport – and certainly not F1.

I had not really thought about automotive as a career before then. Formula Student made me realise that motorsport combined academic and practical aspects, both of which I really liked. On top of which, I found I enjoyed the element of teamwork. Even so, Formula 1 seemed a world away from whatever I might choose as a career.

But during my final year at Queen's, McLaren Racing sent an advertisement directly to universities, saying they were looking to take on two newly qualified mechanical or aerodynamic engineers for their graduate scheme. Professor John Orr, our lecturer in Formula Student, said everyone should be applying for this type thing. I gave it some thought, but didn't do anything. When John discovered this, he actively encouraged me to apply. So I did, almost as an after-thought.

When I got through the first round of interviews, there was an online assessment, which featured mental tests to check, I guess, how your brain worked. Mine must have been okay because Ed Thompson, another member of our Formula Student team, and I were invited to join ten others in a further assessment at the McLaren Technical Centre in Woking, Surrey. This involved working in a group and also doing something on your own. I remember thinking: 'Well, if nothing else, I've seen inside this famous team with the glass lifts, the boulevard with its impressive collection of race-winning cars, and the massive trophy cabinet.' It was quite an experience.

Both Ed and I were called back for a final interview. Ed was offered a place with McLaren Automotive, where he still remains. I was interviewed by the technical director, Paddy

Lowe, and Jonathan Neale, the chief operating officer. The three of us sat in a room for what seemed like no more than a pleasant chat over tea and coffee. I then faced a theoretical engineering problem posed by one of the engineering team, designed to test my use of first principles of engineering and give an understanding of how I would work through a problem. It was difficult; the hardest thing I'd ever had to do.

When I got back to uni and chatted with my lecturer, he said the reason for the tea and coffee and cosy atmosphere had been to get me relaxed enough to say something I was not intending to say! That had never dawned on me. I was unsure what the final verdict might be.

All of this had taken place before the final year Formula Student event at Silverstone. On the Saturday night, I got an email to say I'd been accepted by McLaren. I didn't tell anyone until we had finished the race on the Sunday. That's when it hit home. Like, 'Wow! This is happening. I'm going to work for the McLaren Formula 1 team.'

Chapter 2

ENGINEER TO STRATEGIST

Starting at McLaren Racing was a big deal in every sense. I had been away from home when studying at Queen's University in Belfast, but I had not spent any time living outside Northern Ireland. Naturally, I had a few reservations, but I knew this was a fantastic opportunity as I moved to Surrey in September 2009.

The first year was a graduate programme which involved working in various departments. I spent time in design, vehicle dynamics and systems engineering; three months in each. The final quarter of the year was spread across the car build, wind tunnels, planning and the machine shop. This was useful because I got to know people in each department. Not only could I walk through, say, the workshop and be able to talk to someone about a specific item, but I also gained a bit of an understanding about the small things that might niggle or irritate each department.

At the end of my first year, I moved to design full time,

starting in engine systems before moving on to suspension. That was great because I worked with guys who had been there a long time and were totally expert in what they did. Nothing was too much trouble. They found the time to explain and go through stuff; it was incredibly helpful.

The good thing about being a young engineer in Formula 1 was benefiting from a fast feedback loop. You would design something, such as a toe-link on the rear suspension, and, within two or three weeks, it was on a car. More or less straightaway you would find out if everything worked as it should, and if your design was good. I really appreciated such strong and immediate feedback.

After two years, I moved to the transmission department. If I ever went back to a factory job, I would probably choose gearbox because it has the perfect blend of academic and practical. I really loved it there. It was a mix of very detailed design, followed by going to the workshop and seeing the gearbox being built before watching a run on the dyno. For me, this was an ideal combination within a little standalone group. I learned a great deal.

Throughout all this, however, I had a burning desire to be trackside, give that a go. During my time in gearbox, I started to volunteer for weekend race support. Usually on the Friday of a grand prix weekend, I would go into the team's mission control (which I'll cover in more detail later) within the factory. There were loads of tasks a junior vehicle dynamicist could carry out to help the race team at the track. I would do things such as handling the weather reports or checking that suspension components were not being overloaded. I worked in mission control for five or six Fridays during a season and my gearbox manager, James Manning, was very good about allowing me to make up the time spent there.

Back then the company was building and selling the new McLaren GT track car. Customers needed a lot of support – as they do with every new car – and I got involved with a programme set up by Mark Williams, an ex-race engineer. It meant going to circuits such as Donington and Brands Hatch with a GT, and plugging my laptop in to make sure the car was running okay. This was my first feel for life in a race team. It was a whole new world. It was also when I first met Ryan, my better half, who was working as a mechanic with McLaren GT.

My F1 trackside involvement began with a test session at Idiada. I'd never heard of Idiada, nor knew that the test track was about 40 miles (65 kilometres) south-west of Barcelona. To give junior engineers the chance to go trackside, McLaren would alternate the reliability engineer at the races. I soon took my turn at the 2012 German Grand Prix at Hockenheim.

This experience would lead to becoming a trackside performance engineer with the McLaren F1 team. Having then shadowed Tom Stallard (now McLaren's race engineer for Oscar Piastri) for two races, I stood in for Tom when he went on paternity leave for two subsequent races (India and

Abu Dhabi) in 2013. As my first serious experience with trackside working, I found it difficult to find my space, even though the system is very clearly defined at the race track and everyone has their role and knows exactly what they're doing.

I found it quite tense to be standing in the garage, trying to make the correct call on things, determining what you need to say and choosing when to get your point across at the right moment. It is mainly about confidence and understanding what is going on around you. All of this was a very useful prelude to doing a full season in 2014 when Tom stayed at the factory and I worked as performance engineer for Jenson Button.

Jenson had a lot of experience. Having been in F1 for nine years before winning the World Championship with Brawn in 2009, and then joining McLaren, Jenson knew exactly what he wanted from the car. He didn't need coaching or the sort of support a younger, less experienced driver might require. Jenson understood how the car would react to certain changes and he had the ability to explain very clearly what he wanted. He was very easy to work with. In addition, Jenson had been with the same race engineer, Dave Robson, since joining McLaren in 2010. Working with them both was an ideal situation for me as someone new to the role of performance engineer. I couldn't have wished for a better introduction. (A fuller description of the various engineering roles will follow later in the book.)

If you look at the results, you will see we claimed one podium: third place in Australia at the beginning of the 19-race season. But we never actually got to see Jenson spraying champagne that day. Photographs of the podium show the winner, Mercedes's Nico Rosberg, with Daniel Ricciardo of Red Bull second and Kevin Magnussen taking third for McLaren on his F1 debut. Sometime later, Ricciardo was disqualified for a fuel irregularity, and everyone shuffled forward one place, allowing Jenson to move up to third from fourth on the road.

At the end of the year, McLaren told me I would not be racing as a performance engineer in 2015. They offered me a number of roles at the factory, but I felt I had worked hard enough to warrant being at the track. It seemed a really disappointing setback in my career. I decided it was time to move on.

There was an opening at Force India for someone capable of doing a mix of performance engineering and strategy. It was supposed to be factory-based but I knew that Force India, being much smaller than McLaren, it would probably lead to a role in the race team. I was aware that Force India was struggling financially, but Ryan, working there as a mechanic, had always been paid. Since there was a strong chance of working on the race team, I thought it was worth a go. I applied and got the job.

The timing turned out to be opportune. As I was moving

from Woking to Silverstone, Randy Singh, the strategist for Force India, was going in the opposite direction. A role that was supposed to be half performance and half strategy suddenly became all strategy – which was an interesting development for someone who was not a strategist! Okay, when I was at the races with McLaren, I had chatted to their strategy team and become aware of some of the decisions they had been making. I had also tried to learn more just before leaving when I saw that the Force India job offer included some strategy work. But this latest development was taking it to another level.

That said, Force India did not have many options. I think the feeling on both sides was, 'Let's give it a try!' In any case, I thought it would be good experience in the short term, until Force India found a strategy engineer, at which point I could go back to doing performance engineering. In the meantime, I had a lot to learn. Because of such short notice on Randy's side, we only had two races together. For the first, I sat and watched how he used software, formed tyre models and how strategy plans and reports were created. For the second, Randy let me do as much as I could while he kept an eye on me. By my third race with the team, the 2015 Canadian Grand Prix, Randy was gone.

I relied quite heavily on management support from Force India: Tom McCullough, in particular, plus help from Oli Knighton, who had been working with Randy at Force

India for some time. My previous experience with McLaren stood me in good stead because I understood both performance and trackside engineering as well as having done race engineering with McLaren GT. I was fully up to speed on such things as the use of the intercom and the importance of engine modes in qualifying. It may not have been a recommended route to strategy – to say the least! – but it was fine. We simply got on with it.

Having seen Force India in the pit lane, I had some knowledge of how they operated and knew moving there would be quite a step change. The McLaren factory is massive and spotless; a very different space compared to the Force India factory and its comparatively small workforce. There were around ten of us when I worked in suspension design at McLaren. When I walked into the Force India office and asked about the location of suspension design, they pointed to one guy. He was Force India suspension design. It was a totally different environment. I remember thinking that this little team, with a lot fewer people and a canteen with a few plastic chairs (as opposed to McLaren's 'restaurant'), was producing F1 cars that were not that dissimilar on the race track to McLaren's. The situation has moved on for both teams since then.

At the time, Force India had a very friendly atmosphere. People simply got their heads down and worked hard. Strategy was a prime example. After just six years in F1, I found myself in a role I would never have been given with another team; they simply would have employed someone with more experience.

Moving to Force India also made me realise how stressed I had become at McLaren without knowing it. I was much happier at Force India, where everyone worked on the same page and just cracked on to get the job done. It was a bit less political. There was none of this: 'Oh, that's not my job.' At Force India, the attitude was: 'Right, what needs to be done next? Okay, let's do that. Do you need help with this or are you okay?' It was a very different environment; it had to be because of such a huge difference in size.

Saying all that, there were certain aspects of McLaren that had really suited me. I like a neat environment and details such as having a clean desktop at the end of the day. That was the ethic at McLaren, but I don't know whether this method of working shaped me in certain ways, or I was always like that. It meant I had to deal with a culture change at Force India, such as stopping myself from asking why, for example, someone had hung a jacket on the back of their chair. Surely that's not what you're supposed to do?

It worked both ways. If you were working late in the design office at McLaren, dinner was available in the evening. Force India did not have the resources to provide anything like that. The flights and hotels used by McLaren were better; certain small things were nicer. The plus side at Force India involved having fun in what was effectively a family environment. Everyone looked out for each other.

Although I had initially been very disappointed at being told I would not be racing at McLaren, it soon became apparent that the move to Force India was actually a good thing. I could have stayed another few years – maybe more – at McLaren without appreciating how unhappy I was. This was a powerful realisation.

Meanwhile, life at Force India was not without its ups and downs. There were a growing number of rumours about the team's funding and its future. Despite this, I was always paid on time; my manager, Tom, worked hard to make that happen. Nonetheless, during the 2018 season reports of the team's possible demise grew in strength. When we got to Hungaroring for the Hungarian Grand Prix at the end of July, team members were called to the front of the garage and told Force India had gone into administration. Despite the growing uncertainty in the previous weeks, this was a shock. It was not new in F1, of course. Teams in the past had gone through similar uncertainties when suppliers could not be paid and companies such as Pirelli withheld their tyres. It was a difficult situation. Despite being very worried about the future, team members accepted they needed to crack on with the rest of the race weekend.

The summer shutdown followed the Hungarian race. We went into the two-week break unsure about what was going to happen on the other side. Fortunately, the assets were bought by a consortium led by Lawrence Stroll, the team was renamed Racing Point – and we carried on. There were certain factors such as new uniforms and going back to zero points in the championship. But, from an operational point of view, aside from the name change, there was very little impact day-to-day.

At the end of the year, it was a confirmed that Sergio 'Checo' Pérez would be joined by Lance Stroll for 2019. It was a mixed season: we finished seventh in the championship. The following year was very good in parts, but not so good when we had points deducted following a protest over the legality of certain parts on the car. And then Checo gave us our first win in the Sakhir Grand Prix in Bahrain (I'll cover that huge milestone in more detail later). It seemed appropriate that the team should then take the beginning of a huge step forward when rebranded as Aston Martin.

After so many years of struggling, this felt like success. There were changes in every direction. In the past, when you said you worked for Force India or Racing Point, it would mean little to anyone unfamiliar with F1. The Aston Martin name was instantly recognisable and felt very positive. There was a new motorhome, upgraded pieces of important kit, and new strategy software. Money was suddenly trickling down to all the right places. This was very different from when I joined. It probably really annoyed Tom McCullough, but I used to send an email every week outlining things I felt we could improve upon. I knew that Tom would be unable to achieve most of the points, but I always thought it was good that he should have the list of things we could target if we had more resources. Now Tom could start ticking those boxes.

Over the years, the strategy group gradually grew from Oli and me to six in total. It was a similar story for other departments, everyone enjoying the fact that there could be all-round improvements rather than struggling to make it through with limited resources.

But along with that investment came a lot more pressure and a lot more expectation, which diluted some of the fun element that had been so strong within Force India. Then Covid happened. From initially thinking only the Chinese Grand Prix would be cancelled, to wondering if there would be any races at all, we suddenly found more and more races being added across the European summer. My job felt secure in one way, yet uncertainty hovered in the background.

There was a lot going on. With the triple-headers, different time zones and all the associated travel, I reached the point where I needed some races off. Several F1 teams now ensure their mechanics can miss two or three races of their choosing, but I wasn't allowed that flexibility. Something had to give. In January 2022 I handed in my notice. The Hungarian Grand Prix in August would be my last with Aston Martin.

Chapter 3

THE NAME OF
THE GAME

Formula 1 is unique as a sport. Unlike, say, football, where you have one team competing against another for a single prize, F1 has two drivers in each team, both wanting to beat each other and win the same championship. At the same time, their team is trying to win the constructors' title. I find that an interesting dynamic within all ten F1 teams as they travel around the world to race in more than 20 grands prix.

Considering the amount of rain we had at some stages during many of the race weekends in 2023, it may seem a contradiction to say that the F1 calendar tends to follow the sun. During the first few months of the season, we visit the Middle East, Australia and the Far East. The middle of the year is spent predominantly in Europe, before setting off to flyaway events across the Atlantic and then returning to the Middle East for the final rounds. At which stage, championship points are added up for the last time and each team takes stock of how much they did – or did not – achieve during a full-on nine months.

A strategist is no different to anyone else on the payroll: you are working for the team. Your bonus is set by the team's finishing position in the Constructors' Championship and yet, within that, there has been this sometimes very tense battle between two drivers looking to establish the in-house pecking order.

It's an interesting conundrum for a team management that wants two drivers capable of regularly scoring points. And it goes beyond having at least one driver doing the business if the other crashes or has a problem of some sort. If both drivers achieve the ideal scenario and reach the top-ten shoot-out in the final part of qualifying (Q3), they can compare lap times and learn from each other. One driver might be strong in one particular corner while his teammate is quicker somewhere else. The team can also develop as a whole if one driver is good in the wet and the other is marginally faster in the dry.

Saying that, having two very closely matched drivers can bring potential conflict for the strategist during the race. If they are running in close company, which of the two do you bring in first for the scheduled pit stops, or in the event of the appearance of a safety car? At times like that, it often struck me that the strategist's job would be much easier if you had a single-car entry. Or, given that there are two cars per team, having two pit crews with individual pit-stop boxes would allow both drivers to come in at the same time and do away with the tricky problem of choosing one over the other.

This is just one example of F1 teams having two halves with individual goals operating within a group fundamentally driven by a single mission. Even allowing for one or two team members being central to both cars, there is this strange paradox of each half of the garage focussed completely on one car for 'their' driver and yet, when it comes to a pit stop, the crew unites 100 per cent to have whichever driver under way in the fastest possible time. Similarly, in the event of an accident and one car needing repairs in a hurry, both sides of the garage will pull together. Everyone then has the same goal. But that dynamic changes once the car is back on track!

There are many elements that go into winning a grand prix. Having a competitive and strong car is obviously one of them. Each driver not only needs to be quick but also reliable and capable of sticking to whatever strategy plan has been put in place. Everyone needs to work together as a team. And, ultimately, the strategy must be effective.

Thinking of the win with Checo Pérez and Racing Point in Bahrain, we maybe did not have the best car, we probably did not have the best driver, but we took advantage of errors made by others. Even if some of the components are slightly weaker than is necessary to win, there is always an opportunity if you make the most of what is available. On that day in 2020, the Racing Point was very strong: Checo liked the Sakhir track; the strategy was good. A lot of the elements came together and then we took our chance when collisions and incidents affected the competition.

But this was not a case of rolling the dice or taking a lucky gamble. It was about doing the absolute best with what you've got and being in a position to pick up what others leave on the table. That can be a very difficult policy to consistently abide by. When you are a lower team, it is tempting to gamble on, say, a perfectly timed safety car in the hope of achieving a lucky result. However, that is not going to work nine times out of ten. It is better to be in the singular best finishing position.

We always opted for the best strategy rather than risk a one-in-a-hundred tactic that might come good. In a bad year you need to be thinking of the long game: building and shaping the race strategy team; streamlining the pit stops; putting whatever is necessary in place and being able to make maximum use of it the following season when, hopefully, the car comes good.

My worry was getting into the easy habit of taking a gamble for the sake of it and forgetting how to put in place an efficient, points-scoring strategy. At the other end of the scale, it would be like Red Bull, having won the championship multiple times, choosing to do something off the wall and losing sight of strategy's main purpose, which is extracting the maximum from whatever you have been presented with throughout any given race weekend. The drivers will obviously be trying to go as fast as they can; the designer will have done everything to present a quick car; the engineers will extract the best setup. The strategist will then use this combination as part of a plan to achieve the maximum number of points for drivers and team.

You could have one driver in sixth place, and the other driver out of the points. Or they could possibly be seventh and eighth. If it is a better points combination, then you should sacrifice the sixth-place finish. Rarely could you engineer something like that, but it is an illustration of working towards gaining the maximum number of points.

Taking the points-chasing scenario even further, you could be competing directly with another team for championship position going into the final races of the season. Even if your team is unlikely to claim points, there could be ways of preventing your rival from scoring points by perhaps creating traffic or closing down their pit window and forcing their driver to run longer than planned.

I remember the 2018 Hungarian Grand Prix when we held up Renault long enough to allow McLaren to beat them. We felt having McLaren score more points was not going to affect our position – but this proved wrong at the end of the year when we finished only ten points behind McLaren. That was a classic example of being solely focused on championship points.

Such strategy does not apply at the beginning of the

season because you don't know at that stage who your direct competitors might be. The first half of the year is all about putting as many points as you can on the board. Only towards the end do you start to think differently. This sort of tactical play comes from being reactive during the race, but there is much that the strategist can do to maximise your chances before the grand prix has started.

During the race itself there will inevitably be variables, such as the timing of pit stops and choosing the right tyres. In fact, it is fair to say that the strategist is continually checking all the angles – tyre degradation, weather, problems with the car, incidents, possible safety car deployment – and asking on every single lap: 'Do we pit or not?' It is easy to think that, for the best part of a one-stop race, the strategist simply sits on their hands. That is far from the case.

Chapter 4

MAKING AND MANAGING

A Formula 1 team is a big organisation, numbering from around 300 people to more than 1,000. Most of them spend all of their time at the factory and never go to the race track. As in any type of manufacturing, there are different departments: mechanical design, aerodynamics, vehicle dynamics, systems engineering, software and IT, manufacturing and build, testing and inspection, marketing, finance and logistics. And, of course, trackside functions, including lots of engineers and mechanics building the cars and travelling to the races.

I have tended to live my life in a small section of the latter because the number of technical personnel in each race team is limited to 58 at the track. In addition to that, there will be people from management, catering and marketing, but 58 technical people are responsible for getting the two cars ready to race and then running the grand prix. The 58 at the track do spend some time back at base in between races but, largely, they are in this isolated bubble and live a very different life to colleagues back at the factory.

Like the race team, no one at the factory has a nine-to-five job. Everyone puts in long hours and works very hard. The really busy period starts in September, with the design for the next year's car, followed by the manufacturing department being full-on through December and into January, often working through weekends with only a brief shutdown for Christmas. It is the one period of the year when everyone on the team is under one roof for more than a few days at a time.

As in any walk of life, younger members of a Formula 1 team do not necessarily understand either the influence or importance of having a good manager or team principal. From a distance, senior management seem to sit in an office and not do much, apart from wander around and occasionally chat to people. There is very little interaction with superiors at an early stage but, as your career develops, appreciation grows of the work they put in, from running the team in the factory and at the race track, to finding and supporting sponsors and generally keeping such a vast organisation running as smoothly as possible. They may have a broad knowledge, but good managers will quietly accept that they do not know the fine detail driving the various departments beneath them.

In my experience, the best managers are those who accept that the specialist groups making up an F1 team are expert in their own specific areas, and their decisions and opinions can be trusted. When I was at Force India (through to Aston Martin), the management very rarely got involved in strategy decisions at the pit wall. They were always keen to join in the analysis and feedback after a race on the understanding that, if things had gone wrong, you would learn from your mistakes. An unfortunately timed pit stop, for example, would have been obvious to everyone the moment the safety car appeared a few minutes later. There would be no need for anyone to state the obvious by pointing out it might have been better to wait one more lap. That principle applies throughout the various departments making up an F1 team, from design to logistics.

Design is predominantly split into mechanical and aero-dynamics, each working as a separate entity. In simple terms, the aero department designs the fundamental shapes of the car, including the external surfaces of mechanical components such as wishbones, brake ducts and anything that air flows over or through. The chief designer (the title varies from team to team) may create some essential principles that the car is going to work around. Mechanical design may then decide certain concepts such as wheelbase, track, ride heights or the rake of the car; aero will then take responsibility for all the surfaces. These components are key because aero performance is actually the most crucial thing, particularly when the change of regulations in 2022 laid even greater emphasis on the significance of aerodynamics.

Some teams will have aero divided into global sections, such as the front, middle and rear of the car. One group will be working around the front wing; another, the mid-section, which includes the radiators and, most likely, the floor of the car; the third will deal with the area around the rear wing and rear brake ducts. With the air flowing between all three, these groups need a good understanding across each other, particularly during wind-tunnel tests, when the needs of all three sections come into play.

The mechanical design group will split much more heavily because it is easier to work on sub-assemblies at the same time. A suspension design group might be separated front and rear. The suspension group would be responsible for what is known as the unsprung mass of the car (everything that, in simple terms, is connecting the wheels to the chassis). For teams that have one, there is a gearbox design group, potentially doing all the transmission as far as the rear wheels of the car.

The chassis design section is the central point – from designing the entire cockpit, to the mounting points – for everything (suspension, radiators, electronic boxes) that runs off the car. This includes the engine which, even for the teams that have a partnership with a power-unit manufacturer, will require chassis design to take care of the engine-mounting points. Teams that manufacture their own engines will obviously have a closer relationship with the engine design, whereas customer teams will have a more difficult job mounting an engine not specifically designed for their car. For example, the Mercedes unit would have been built specifically to go in the back of their own car.

During my time at McLaren, we were taking the Mercedes engine more or less as it came and working the McLaren chassis around it. Similarly, when I did gearbox design at McLaren, we were working on the Mercedes engine, with its particular pickup point for the oil pump and the mounting points for suspension, and whatever else. The FIA technical regulations did eventually standardise the mounting points between the engine and the gearbox, but the installation of the engine still presented its difficulties because, for example, the oil pump did not have a standard pickup point, and neither did some of the other mounting locations. So the mandated six-bolt pattern seemed a bit arbitrary. Most teams also have an engine systems department dealing with radiators, coolers, and anything that wasn't supplied by the engine manufacturer.

Compromise plays a big part, which is when the chief designer gets involved. It is said that Adrian Newey is very effective at Red Bull because he pushes the mechanical side to prioritise the aero setup. Everything is about give and take: what is the biggest priority? What is going to make the most difference?

Underscoring all this, everyone will want their piece to be perfect. Mechanical design will call for the biggest range of cooling possible. Aero will want the engine cooling to be as small as it realistically can be. This is where you need a chief designer with a good knowledge of what can or cannot be done to level everything out. Despite everyone fighting their corner, the various groups are generally onboard with the need to work together.

In the early design stages, the specification of the car will not be an attempt to cover every possible race with, say, the maximum temperature ever seen. It is accepted that some engine management will be needed at times because the car would be massively over-designed if the highest temperatures were considered. Similarly, certain components such as suspension members will not be designed with the hardest side impact in mind. These parts should be able to withstand a percentage of the load because, otherwise, you would have the heaviest car on the grid. Saying that, teams have been known to use stronger components for tracks such as Monaco.

The fundamentals are set out early on, followed by detail such as the aero group saying: 'Right, this is where we want the radiator to fit' and mechanical responding: 'We can't quite make that space work.' The subject will go around the loop again and again in meetings where the chief designer will oversee and rule on such interactions. The chief designer, by the way, might alternate from year to year, allowing one designer to focus on the current car while the other chief designer is looking a year ahead – but both working in parallel.

All told, there would be several meetings in a similar vein. Mechanical might be saying, 'We can't possibly design this piece of suspension within the aero surface,' thus setting in motion another loop of discussion and debate. The person responsible for designing the rear suspension would go directly to his or her counterpart in gearbox and jointly figure out where a certain bracket or mounting point should be located.

Within all this are sub-sections doing detailed aero design for, as an example, bodywork and front wings. Aero will provide their requirements for suspension surfaces, while the mechanical group, using CAD (Computer Aided Design), will design the suspension piece strong enough to do what is required, but within the shape set out by aero. Quite often, there will be a certain amount of back and forth between the two departments should theoretically the best shape not quite work from a mechanical point of view. Although separate within the engineering office, these groups are working very closely together. Within each, there will be support groups. Stress analysis will check some of the design parts to ensure they are strong enough or, perhaps, to see if they can be lighter.

At the same time, there will be checks to establish if the parts can survive the worst-case bumps or side impacts. A materials department will be thinking about the effect of temperature. A carbon fibre expert will advise on the best type of laminate to resist certain temperatures or loads.

When working on 3D models, it pays to go across to other sections of design and ensure a part that should be loaded in the middle of your piece has not been missed.

So, there is a lot of interaction between departments, and I enjoyed that element of design work. When working in the gearbox design for two and a half years at McLaren, part of this experience meant trying to get everything gearbox related inside a predetermined aero surface. I also developed the oil system, including fluid flow considerations; another element which I really liked.

Some components, particularly those that are specialist or require labour-intensive machining, are manufactured outside the company. Otherwise, parts are made in the team's machine shop, which is usually within the lower ground floors of the factory and often directly below the design office. It was useful to be able to go downstairs and see a particular component being made on a machine. It also worked the other way, in that the machinist could come to design and ask a question if an element of the drawing did not make sense or, perhaps, a discussion could be opened up about how a small adjustment on the drawing would make the part easier to manufacture. From the machine shop, mechanical components might go through treatment processes, but will certainly undergo inspection before being sent to either the stores or car build.

Ultimately, everything will go to the chief designer and he will assess the compromises that have had to be made across the board. The chief designer might not be involved in the detailed design at any stage but he is responsible for how it comes together, ensuring there are no massive clashes or errors. He will make the big calls when it comes to having to choose between the needs of one department against another. And, as we have seen, these are many and varied!

Most teams have a department focused solely on software. A strategist's calculations depend on the mathematical model put in just as much as the capability of the software. For example, when establishing a tyre model, the accuracy will depend on the estimated laps of tyre life programmed into the model rather than how the software functions. Several teams use the same off-the-shelf software, as well as adding bespoke pieces of software that they have designed and developed.

The need to have data quickly drives a lot of the database development. You might want information on, say, the last wet race at Monza, and the bespoke database will take you straight to it. It is all about having details data-based: 'When was the last time we had a fault when the wheel broke?' That sort of thing.

A lot of software is built around the team's infrastructure and how data is shared across groups. For example, when looking at a Free Practice 1 (FP1) session and comparing two components, how did the strategy group input the track improvement? How is the tyre group registering the difference between the two tyres? How all these elements work together is one of the biggest things for software to handle because more data than ever before is being generated. It is important not to confuse a result with something you did not know was happening at the same time: it is very easy to look at two runs and say the second is 1.5 seconds faster – but ignore the fact that the driver made a mistake during the first run. It is about trying to isolate the effects of each factor; software can do a lot of that.

Like any business, an F1 team has the usual functions of HR, finance, logistics and communications. The one major difference is that some of these departments cannot work in the way they would expect in, shall we say, the more traditional workplace. With IT, for example, the accepted routine elsewhere would be to log the fault and wait for someone to come to your desk, fix the problem or, perhaps, arrange for a new computer. With the time pressure inherent in F1, it simply cannot work like that. There is no time to wait for help because the car is hitting the track right now! IT may be brilliant at what they do but Industry Best Standard, good enough elsewhere, simply does not apply in this environment.

Marketing also has a very important role to play in the survival of the team, even if, as a strategist, you have little connection with the promotions group during the race weekend. It is common to find several important guests or sponsors in the hospitality area, or on a tour of the garage. This can make for an interesting scenario if you are rushing into the garage on an urgent mission and come across a group of VIPs being shown around by the marketing department. The immediate dilemma lies in deciding whether to politely push your way through, in what is usually a busy environment. You need to assess if your mission is more important than the value the guests might bring to the team, either now as a commercial partner or a potential sponsor in the future. Tact and diplomacy is usually the best solution even though, at the time, it may feel the team's race performance depends on the successful resolution of whatever your pressing problem may be.

Some races bring very few pit visitors, but it will come as no surprise to learn that Monaco is the worst case: you can't move for people in the already cramped surroundings. Occasionally, these visits coincide with pit-stop practice. I always found it strange that guests sometimes stand very close to the action despite pit-crew members and the associated wheels and equipment travelling in all directions. I can't think of any other top-level global sport where such close contact is allowed.

A pet hate of mine was having lunch in hospitality when a complete stranger, usually drinking from a glass of champagne, would come across and begin asking questions. Melbourne could be difficult because, being the first race of the year, there would be an element of panic and you would be preoccupied by having to deal with many unknowns.

Occasionally, I was asked to speak to a group of guests if

they were particularly interested in the pit wall or strategy. They might represent a new IT supplier or partner, in which case I would focus mainly on the strategy software. Along with everyone else in the team, judgement is clouded by the belief that your role is vital. Meeting guests and commercial partners in this way does reinforce the fact that, without significant investment, none of what we do would happen.

The same thought can be indirectly applied to the role of the communications department in gaining publicity for the team. But, then again, the thoughts of the trackside technical team would be coloured by drivers invariably turning up late for meetings because of – allegedly – being delayed by media work! My involvement with the communications team tended to be limited to checking the post-race press release if mention had been made of strategy, our pit stops and why we chose a particular tyre. It might be the same for qualifying, particularly if something had gone wrong. The press officer would check if anything had been missed or didn't ring true in what was being said in the release. Very rarely would I be asked for a quote; my role was checking for accuracy if strategy was mentioned.

It is always useful, however, to read driver quotes on press releases issued by rival teams; there is always something to learn. Through experience, you become familiar with the thinking of various drivers. With some, if they say they are going to do a two-stop, you can be certain the opposite will happen!

The question of a team's proposed tactics is always the subject of speculation by the media. I was never asked such a direct question by anyone from the press – mainly because they knew it would be a waste of time! But there would often be a more subtle approach: a journalist might have just spoken to a rival, who said they were going to stop once – what was my opinion? There were times when it was necessary to think about your reply. Each team has strict protocols in the event of an incident – great or small. There would be clear demarcation over who was allowed to speak to the media and who should give feedback on whatever had happened. Thankfully, I was not one of those spokespersons.

Since working as an F1 analyst, bringing my strategy knowledge to Sky Sports F1 TV coverage, I have a better appreciation of how the media has to work with very limited data. I no longer have a method of working out pit loss, for example, and need to rely a lot on what others tell me. Having never had the need to think about it before, I now appreciate how the media works. Like everyone else in F1, they have a particular role to play.

Communication with the media has improved over the years, teams generally becoming more proactive. Pirelli run a WhatsApp group, giving tyre information and listing the various sets drivers have left to play with. Teams will outline some strategy thinking: mainly generic topics such as the choice being between two and three stops, but without, of course, getting into the exact detail of what they're going to do.

Inevitably, I will have come up with my own thoughts on strategy – sometimes the team briefing will fulfil that, and sometimes it won't. The teams will usually give pit loss, rounded to the nearest second. That was something I would do when asked. By rounding pit-loss time, (the time taken to enter the pit lane and complete a pit stop relative to staying on track), it would be sufficient for the media's needs, but without giving away the fine detail of what we were hoping to do. I always tried to be helpful.

Of course, with my background, I am now asked to go along the pit lane and see what I can discover – but I find I'm terrible at that. I hate asking people what they know, probably because I empathise with them too much. I am hoping that will wear off over time because I do want to ask the tough questions!

Chapter 5

GOING WITH THE FLOW

Like everything else, wind-tunnel development has moved on a huge amount during the comparatively short time I have been in Formula 1. When I joined McLaren in 2009, they had their own working wind tunnel, which was exciting to see. The McLaren tunnel was built for a scale model, but an increase in the size of cars and their wings brought interference from the walls and the roof of the wind tunnel, rendering it unusable. Many of the teams then found they had little option but to travel to Germany to use the Toyota facility with its full-size tunnel.

Aero design can be tested using two methods: CFD (computational fluid dynamics) and a wind tunnel. If mechanical design has off-line simulations and a simulator, then aero design is followed by CFD simulating the flow of air, and then the wind tunnel. It is the same three steps for each, with the wind tunnel predominantly working as an aero tool. Being a good aerodynamicist is probably one of the most difficult jobs in the pit lane because everything on the car relies on how good the aero is. It is fundamental to everything we do.

The number of CFD runs (which are effectively computing air flow, downforce and drag) is regulated by the FIA (the Fédération Internationale de l'Automobile; the sport's governing body) and checked through the amount of computer power used. Because aero is so important it is subject to more regulatory control than mechanical design. For instance, there is not a limit on simulator runs, but there is a restriction on CFD and wind-tunnel use.

The wind tunnel, like the simulator, brings in external effects. Various grades of roughness on the wind-tunnel belt will simulate an abrasive road surface and the boundary layer off the floor from the wind. The model in the tunnel can be turned 30 degrees relative to the prevailing wind, which mimics a through-the-corner situation. (The car is turned only one way on the assumption that the opposite way is symmetrical.)

The time taken to position the model is critical when it comes to measuring wind-tunnel operation. The FIA measures 'wind-on' rather than 'tunnel-on'. Let's take using a vacuum cleaner as an example. Having finished in the hall and moved the Hoover to the floor of the dining room, if the Hoover stayed on, all of that would count as 'wind-on'. But if you switched off the power, that would not count and would therefore be more efficient in 'wind-running' terms.

Extending this principle to the wind tunnel, it becomes important to ensure the model is positioned correctly, with the road running and everything, literally, up to speed before hitting the 'wind-on' button. The same thinking applies to finishing the run. If it becomes apparent early on that the test component is not good enough, then cut the wind immediately before doing whatever else is necessary to bring the process to an end.

There is a model design and build department within the factory devoted to working on 60 per cent scale car models that can be used for testing. People assume this must be boring because the model makers never see the real car. In fact, it's very interesting because you see components before anyone else and, having monitored the passage of air moving over them, quickly establish whether or not the parts are going to work as hoped.

Downforce can be measured through load sensors placed on the suspension, with the load being evaluated in many ways. It is about trying to have a grand scheme of how everything comes together. On a brake duct, for example, you might be testing the cooling properties as well the passage of air around it. In general, the greater the increase in cooling, the more negative the effect on downforce. But if the duct is efficient in terms of cooling, then the accompanying loss of downforce is an acceptable trade-off.

The movement of air is measured by sensors. Cars will be seen carrying large rakes – which look huge grill trays –

during testing or, occasionally, for the opening laps of a free practice session. These rakes consist of pitot tubes – little tubes measuring pressure drops relayed to a sensor buried within the car. It is also possible in the wind tunnel to dye the air, which shows up the effected air flow seen with the aid of a UV light.

Flow-vis paint can indicate the passage of air in the same way, and is often seen being used on the actual car during test and practice sessions on track. As a coloured paint, flow-vis will normally be visible to the naked eye. When used on track or in a wind tunnel, the passage of air on flow-vis will have the same effect as rain being blown in straight lines towards the top of a windscreen. When it comes to the actual race car, the aero department will be looking for these flows to replicate all they have seen on the model or in CFD. Above all, they will want to avoid air becoming detached from the wing and causing the car to stall because downforce is absent. Although flow-vis is low technology, in these circumstances it is an unintrusive way of getting a clear picture. There are occasions when, with the best will in the world, components will not respond on track as they did in either the wind tunnel or CFD.

Wind-tunnel time has been limited in recent years, the amount of restriction based on a team's championship position. Basically, the more successful a team, the less time allowed for development. FIA regulation requires scale models to be used, and more teams are working towards having their own wind tunnel on site. It can be detrimental to build a model and move it some distance for testing in someone else's wind tunnel. You can never be sure of the tunnel's condition, or when it was last calibrated. Having your own facility means switching it on and off whenever you like – although building and running a full-size wind tunnel is a massive expense.

A few of the departments within the factory are much more theoretical than others. Vehicle dynamics is one of them, in that they propose systems which, in fact, may not be physically possible. They might put forward a suspension system that, physically, no spring or damper could possibly achieve. Their aim is to show a car characteristic they want to realise in a perfect world. Having found a mathematical way of modelling it, vehicle dynamics will then leave it to the next department to try to find a physical way of making it. Vehicle dynamics is the study of vehicle motions: how the car reacts to internal or external movements, driver input, the track, environment and engine. They are in effect saying, 'If we had this reaction, it would be ideal for us.'

In broad terms, the aim is to achieve a car that the driver can take as close to the limit as possible. We talk about 'on the limit' as being on the limit of grip. If you drive slowly around a corner, you will be in total control because there is plenty of grip between the tyre and the road surface. By taking the same corner with gradual increments of speed, you will reach a point where you can no longer make the corner: the tyres lose grip and the car begins to slide. The trick is knowing at which point the speed or the angle is such that the tyre has got just enough grip to make it, and the car is not going to break away. In an ideal world the perfect car would have the driver at that limit – be it cornering or braking or accelerating – at all times. But never over the limit – or, equally, never under it.

Under acceleration, you are going to be limited by either engine power or the tyres (rather than, say, a poor gearshift). When braking, it is about how hard the driver can hit the pedal without locking the tyres. In the corner itself, what matters is the driver feeling confident enough to go as fast as possible without taking an extra margin of comfort. Any additional margin increases lap time.

The car needs to be predictable. An unpredictable car leads to that extra bit of margin because the driver is concerned the brakes might lock or the tyres will lose grip and cause the back of the car to break away in a sudden snap of oversteer. All of which means time lost. Understanding the limit means keeping all four wheels in as much contact with the road as possible; vehicle dynamics focus on that.

Vehicle dynamics also work closely with the race and trackside engineering departments because they are evolving a development route created by some very bright ideas. They then come up with actual designs for the car, or thoughts on, for example, what the range of springs and dampers should be. This will include theoretical and computer simulation work, trying to answer questions such as: 'How much load do we think would be on the tyres?' Or, 'How much fuel do we think we'd use?' Or, perhaps, more complicated questions such as how hot the brakes might run on a specific track. Some of the most intelligent people in the business operate in that group because their minds work in a very different way. I did spend some time with vehicle dynamics. It is very intense computer work, but really interesting.

Within that division sit other groups that link into race engineering. The tyre group will be doing technical analysis and research into how they think the tyres will behave, or what pressures or temperatures we should run, or the best way to run the brakes to get the most out of the tyres.

Along the way, attention will switch to the simulator, which uses a different model method compared to pure computers in that you have a physical simulator that is predicting how the car will react. Compared to the pure computer model that has a simulated driver reaction, this has a physical driver reaction. The computer will react in one way to, say, a slight drop in grip, but it is interesting to see how the driver will react, which is often quite differently. This is where the physical simulator comes into its own.

The simulator is really useful over a race weekend. The guys and girls running the simulator start their day at around the time the on-track team is preparing for Free Practice 2 (FP2). Using a test driver, they will mirror what we do at the track, the object being to achieve a similar lap time, reaction and feedback by correlating whatever is happening on track.

At the end of FP2, when the race drivers indicate what they would like to improve for the next day's running, the simulator team will work on finding a solution overnight, the aim being to provide the on-track team with a list of suggestions to try in Free Practice 3 (FP3) the following day. It might be softer front suspension, higher rear ride height, rear and front-wing adjustments, suggestions on tyres – a whole host of things.

This does not happen every Friday because the circuit model might not be good enough, or there is not a test driver available. If that is the case, vehicle dynamics will simulate setup changes on their computer in the hope of at least suggesting a development direction if a direct answer cannot be found.

Although the simulator is mathematically modelled very well, it is important to be very careful with the results. At McLaren we always treated the simulator results as a direction rather than the ultimate setup. Adding downforce, for example, would be a good direction – but the simulator model might not necessarily tell you how far to go.

The simulator is very accurate in other ways, however. You can change wind direction, track temperature, and decrease or – as usually happens during a race weekend – increase track grip. If this does not correlate with what happens on track, it begs the question: is the lap time not matching the simulator because the test driver is not quick enough? Or is the wind in the wrong direction, or the track grip incorrect? It is a matter of matching the correct element and knowing exactly which element you are trying to correlate. All of this is very complex, which is why the people working on the simulator are really, really bright!

Chapter 6

READING THE ROAD

The simulator (or 'sim') can be extremely accurate, but not perfect, because of the discrepancies you might expect with modelling in the early stages of car design or, perhaps, before going to an unknown circuit.

In the case of a new venue, track details can be fed into the simulator, but there might be a bump or a kerb or, maybe, camber in a couple of places that turns out to be incorrect in the model. Without realising it, the sim driver could be taking a slightly different line than he would if driving on the circuit. Or we might find, when it comes to the real thing, that a kerb can be used more than anticipated, thus increasing straight-line speed, changing the anticipated lap time and bringing a fundamental change to the car's setup.

Circuits the teams know well, such as Silverstone and Barcelona, are a different story. When changes to suspension through, say, the addition of a new component, are programmed into the simulator, the driver should be able to give reasonably accurate feedback on the effect that has on the car, either positive or negative. You might also discover, for example, that a different ride height is preferable.

There are several ways the grip level can be tuned on the sim to give an idea of how much feedback will come from the tyre. Saying that, the tyre is difficult to model because it is nonlinear in its reaction to a number of factors, such as the roughness of the track surface or the temperature generated by the car sliding. Whereas a spring is linear in the way it responds to load, tyre behaviour is governed by temperature, both internal and external.

The sim engineers can determine the effect on the car of variables such as a change in wind direction. The simulator is a complex piece of kit that requires a specialist team – including the driver – dedicated to ensuring it is as accurate as can be when called upon to perform certain tasks. It is a continuous cycle of trying to ensure precision on the sim. In an ideal world, you want to get the race driver into the sim post-race, set the car up exactly as it was on the Sunday and see if the sim is giving similar responses. Once that is established, you can then focus on the next circuit.

The cut back on testing during the season means there is almost zero opportunity for the simulator driver to get into the actual car on track. Occasionally, the sim driver might get a run during free practice on a race weekend, or perhaps he or she might carry out some tyre testing – but even that is a problem because the testing is usually on unknown tyres and therefore the team does not have an established baseline to work from. This is why teams are very keen to get race drivers into the simulator, even though it can be a struggle because race drivers are generally not keen on spending time in the sim, particularly after a grand prix weekend.

Meanwhile, mission control will be keeping across all of this. Each team has a mission control – although some may use a different name. The need arose when teams were limited (because of cost saving) to the total of 58 working people (mechanics, pit crew, engineers, technicians) at the track. Since the restriction does not apply to team headquarters, it did not take long for an important satellite of the race team to begin to grow in significance at the factory.

Apart from anything else, mission control provides an immediate saving on flights, hotels and whatever else is necessary to operate at the race track. The other important factor is the curfew restricting hours worked at the circuit. Since there is no curfew at mission control, team personnel work all the hours that are needed. Which is convenient when the curfew kicks in at the track on Friday night and the race team can have mission control continue the analysis or whatever needs doing.

Mission control tends to be a theatre-like environment, with seats and worktops in stepped rows facing a video wall or screens carrying everything from onboard footage, lap timing and general information that is available to everyone. There will, of course, be audio and video connections with team members at the track. There can be between 30 and 40 people in mission control, ranging from strategy function, race engineering support, aerodynamic support and vehicle dynamics to design and reliability. Some will be linked by intercom to the simulator, which is usually close by.

Mission control is more or less replicating functions at the track, with the link between the two being in real time. Data from the car goes directly back to base; video feeds are live and every position within mission control will have an intercom panel connecting to the pit wall, or wherever is appropriate for each particular role. A few channels will allow specific groups to talk directly: in strategy, for example, we had our own discussions between track and mission control. Many teams have video links between the two, allowing trackside to see exactly who is in mission control. All told, mission control is a powerful resource.

As with everything, there are advantages and disadvantages. Being a very isolated environment relative to the track, mission control has certain positives. Compared to the noisy and sweaty atmosphere at the pit wall in, for example, Singapore, mission control is comfortable and calm. It allows more concise and clear thought than in what is, literally, the heat of battle at the race track. There is room to breathe in terms of desk, computer and screen space, allowing a more methodical way of working that is often difficult to achieve at the pit wall with its constant stream of inputs, pressures and external noise. But then, anyone in mission control going home to a young family may not get enough sleep, particularly when trying to rest during the day to be ready for working in the evening because of the grand prix being in a different time zone. There are no screaming kids in the hotel room when you are part of the race team!

Looking at it another way, that aura of quiet and calm within mission control can lead to feeling divorced from the fine but important detail of what's going on. It is very difficult, for example, to build an accurate picture of ground conditions at the track. That does not necessarily affect everyone in mission control but, from the strategy point of view when, say, it is wet, you lose the sense of whether it is going to rain harder, or the track is likely to dry quite quickly. Has it become more windy? Is the light fading faster than expected? All those feelings are easier to pick up on the pit wall, and they accentuate the need for an informative link between mission control and whoever is trackside.

Underlying that, the strategist in mission control needs to be aware of the perception and way of thinking of whomever they are communicating with at the track. For example, tending to be pessimistic when predicting the weather, my default position was to assume it would rain if I could see it some distance away on the radar. Others in the group might be reasonably optimistic. So I would know, when they said it was going to rain, it was definitely going to rain.

When I was trackside I would try to build a picture for mission control: 'It's raining, level 2. Only need a light jacket.' Or, 'It's heavy and we're going to need full wets.' Or, 'It's freezing cold.' Anything that informed and helped the decision-making process. Of course, some of this may be clear from the TV or on-board images. But some of it is not, particularly at the start of a session when there are no cars on track and no TV coverage.

When working in mission control, I was surprised to discover how much I missed the routine that comes with being at the track. Some teams do it differently, but our schedule was to be in position on the pit wall 15 minutes before the pit lane opened. So I knew, when I walked out of the garage, stepped onto the pit wall and completed my radio checks, that track action was soon to start. That was the way I worked, and it was a signal that my full focus was on the session ahead. I was subconsciously saying, 'I'm in a different place. I have my headset on. I'm in race mode.'

In mission control, your place of work does not change once the track action starts. You have been sitting at the same desk all morning, perhaps having a cup of tea and going to the bathroom, only to return and realise you have missed the radio check because there has been no change in thought or tempo. There is none of the buzz that comes with the build-up in atmosphere at the race track, much of it created by the surge of anticipation from the grandstand opposite. There seemed to be less pressure in mission control, less focus in that moment, because the only background noise might be the hum of the air conditioner. I found that difficult to get used to. Obviously, when the session or the race starts, you're on the case. It just took me a bit longer to get there when in mission control.

It may sound obvious, but you are much more aware of the bigger picture when stationed on the pit wall. If there has been a problem with one of the cars during practice or qualifying and it is being worked on in the garage, one glance across the pit lane will tell you that, even though time is getting short, the car will not be going anywhere for at least ten minutes. Mission control may not be able to see this and may come across the headsets with a message saying the car needs to be out in two minutes if the programme is to be followed, or there is a suitable gap in traffic on track. Anyone sitting on the pit wall would never say such a thing because they know the full story and the guys working flat out on the car do not need to hear it.

Because there are never enough people to handle all the work, it was important to be very regimented when it came to splitting resources. Within the strategy department at Aston Martin, we would have three people in mission control and, at the track, myself or Pete Hall, senior strategy engineer at the time and my eventual replacement. It was essential for each person to have a clear plan for the different sessions.

There would be defined roles depending on whether the session was wet or dry; each condition bringing different priorities. Similarly, your needs during a practice session would not be the same as in qualifying. Everyone had to have their specific role and needed to be empowered to either pass information up the chain, or ignore it if the facts were felt to be irrelevant.

Mutual trust was essential to make this work. If, for example during a race, someone said to me, 'We need to stop this lap because we're going to be undercut,' I might not have enough time to check all the necessary information. (An 'undercut' being when the driver behind stops earlier than the driver ahead and, either through better car pace or faster fresh tyres, emerges ahead after both have completed their pit stops.) I needed to accept and trust what they were saying because there would be no value in having four people doing the same calculation. Confidence and belief were necessary in such an important environment while dealing with the stream of information.

Teams use different methods, but a common approach is to have one person in mission control looking after each car. At Aston Martin in 2022, one person took responsibility for Sebastian Vettel while another was the link to Lance Stroll. Information relevant to Vettel would be fed direct to his race engineer; it would be a similar story for Stroll.

It could be that Vettel's pit-stop window is lap 27 and his lap times are deteriorating because of high tyre degradation. The strategist on the pit wall will make decisions based on the information coming from the people focused on Sebastian's car rather than having to personally spend time doing the relevant working out. The strategist will have been informed about, say, the lap times of a rival that could undercut Vettel if Sebastian stays out and his lap times continue to fall away, thus helping the strategist reach a quick and correct decision.

This working environment and structure is very important. It means the strategist can train more people and have them watch and learn in mission control. They can work with other departments and pick up information from the person sitting alongside. They might hear about a brake problem, work out how that could affect what happens next and pass that on to the pit wall.

Like everything else in F1, the mission control concept has grown over the years. During my time at McLaren, there were no more than 15 engineers in a compact but very nice amphitheatre. The set-up at Force India was even smaller but gradually expanded during the time I was there. In fact, one of my last projects at Aston Martin was to design and develop a mission control for their new factory. In leaving Aston Martin during 2022, I missed out on enjoying the benefits of a sparkling new mission control with all the proper facilities.

Mission control is also very useful for volunteer support, which is made up of people with factory-based roles who want to experience some form of interaction with the track. They could be in the design office or working in aero. By volunteering for mission control, they can help with anything from monitoring the onboards of rival teams, listening to competitor radio or giving the pit wall feedback if other drivers have an incident or accident.

When at McLaren, I did a volunteer role with vehicle dynamics for a couple of races. Listening to an intercom or watching the garage camera offered experience of how frantic it can be. As an engineer or a designer, when you make a request such as, 'Oh, could you just fit this part and see what it looks like on the car?' you need to appreciate the sense of urgency surrounding the race weekend and come to understand that perhaps your request is actually impossible. Volunteering to work in mission control is valuable because it brings home how full-on it can be at the race track.

Saying that, mission control itself has become a big support operation, with many more engineers than the ten or 15 that travel to the races. The output from mission control is much greater than the workload produced at the track, to such an extent that the trackside strategist must deal with a mass of material coming in from several sources, which can range from analysis of lap times to weather information, video footage of your team's cars and other cars, or relevant excerpts of intercom conversations between rival drivers and engineers. That is a lot of data needing to be examined as closely as possible and acted upon if necessary. Formula 1 has reached the point where the teams rely heavily on support from mission control. That was brought home to everyone at Aston Martin during qualifying for the 2017 British Grand Prix.

Eddie Jones, then England rugby coach, was with us that weekend to learn how we communicated with each other and to see if any lessons were worth transferring to rugby. Eddie had spent Saturday morning at the factory before moving across to the track to observe how we dealt with qualifying. No sooner had Qualifying 1 (Q1) got under way than we lost the connection to mission control. There was nothing coming through. Zero! We were at the pit wall, working without the usual flow of detail about traffic or the weather or lap times. I seem to remember our computers also went down. Meanwhile, 20 cars were on track and qualifying was fully under way. It was chaos – but we survived. I remember counting the number of times a car went past the pit wall to see where we were on the run plan. We had to rely on the strength of several very experienced people on the pit wall. We didn't do everything perfectly, but we did make it through to the next session, at which point communications were up and running again. Then it was a case of trying to fill in as many of the essential blanks as possible, leaving the rest until later.

This was painful proof that, over time, the pit wall has become reliant on the extensive background team. As a strategist, you are looking after two cars, determining run times and where your drivers are on the track. Are we impeding anyone? Are conditions at the absolute best right now? It is complicated on a dry track, never mind one that is wet but starting to dry. Are we going to make it through? Do we need another tyre set at this stage? It's hectic.

This also proved why you need people at the track, because sometimes the complex and carefully organised systems are not as robust as expected. Losing the factory only happened once or twice – but it was a mess each time. It had to happen, of course, when someone such as Eddie Jones was there to see how efficient and effective we were! The whole thing was made even more embarrassing because our factory was just beyond Silverstone's main gate. It was the closest we would ever be to mission control during the season. That failure made you wonder just how we manage when thousands of miles away in China or Australia. But that's what the dedicated IT team is for. It is the same with the intercoms provided by an outside supplier. There is a bunch of people in most departments dedicated to ensuring everything works.

Maintaining the intercoms and the pit-wall links are typical of the many F1 jobs that no one really cares about until the system breaks down when you are in the thick of track action. I've never heard anyone compliment the IT team and say, 'Well done, guys. The pit wall is working.' But there is plenty to be said if the system goes down. When I worked on gearboxes with McLaren, we used to joke that nobody ever said a gearbox won a race or made the car fast. But you would hear reference to 'the gearbox cost us points' if the car retired due to an issue or failure. It is all rather like the strategist, in some respects. Do your job efficiently and correctly and not much is said by those outside the team. But get it wrong and social media is full of comments about the part played by the strategist in losing a race that should have been won.

Chapter 7

TRACKSIDE TASKS

Broadly speaking, the trackside team is split into two: the technical engineers and the mechanics. The chief mechanic is exactly that, and is in charge of the mechanics, who are split evenly between the two cars, each with a number one mechanic, the rest carrying out roles that are mirrored on each car. There will be mechanics concentrating on the front end, with others working mainly on the rear of the car. There will be people focusing on tyres, fuel and electrics. Others will be responsible for looking after the trucks (at the European races) and the environment everyone works in, regardless of where we are in the world.

It is important to say at this stage that each of the jobs outlined above is the mechanic's primary role. Their secondary job is pit stops. F1 teams do not recruit someone just because they have, say, the build or fitness to handle a front-wheel gun, or deal with taking the wheel on or off. The pit-stop role – vitally important though it may be – is over and above what mechanics normally do. It may not be the job they are employed for, but mechanics love pit stops.

Most teams have a similar split on the engineering side. On each car, there is generally one performance engineer and one race engineer. The race engineer is the person you hear speaking on the radio to the driver. They are the leader when it comes to deciding on the car's setup and everything that happens to that car, as well as being the main communication channel with the driver.

Below the race engineer comes a performance engineer. Teams split these two roles differently. In some cases, a race engineer is focused on the physical setup, while the performance engineer deals more with what we call 'the car tools' – things such as brake balance, fuel and maps. These two engineers work very closely to get the best out of the race car and driver.

Race and performance engineers tend not to change from race to race although, with the growing F1 calendar, there has been an increasing tendency to think about rotating some of the core people within each team – some senior strategy engineers now alternate with their counterparts between the factory and the track.

In between races, the race and performance engineers would have various roles at the factory. Some would analyse what had happened during the previous race weekend and liaise with the other departments to try to find answers. Was it, for example, an aerodynamic problem or a vehicle dynamics problem? They could also try to find solutions by spending time with a driver on the simulator.

A systems engineer on each car will take care of things like the launch off the start line and how the car is operating physically. Is everything working as expected? Are all the sensors operating as they should? Are there any faults that need sorting? Different teams define these roles and ties between them in slightly different ways.

There will be a number of tyre engineers at the factory. The tyre engineer travelling to the races is generally across both cars and responsible for giving advice on setup to ensure the tyres are being used efficiently. During the race, the tyre specialist will advise the race engineer how, by driving more efficiently, the driver could get the best out of a particular tyre. All of this information will flow through the race engineer to the driver. I will explain this in more detail later in the book.

There will also be a reliability engineer, checking that new parts come to the car correctly; and inspecting the car after a run for any faults or pieces that are broken. The fault system in F1 is really special and continuously improving; designs are constantly changing in order to try to fix or mitigate little faults, because the reliability has to be so strong.

Overseeing all of the above, a senior engineer will resolve any conflict or decide what parts go to which car, which driver will run a new front wing, or which driver will test, say, the hard tyre.

The team principal has overall responsibility and ensures

all the functions work together. A sporting director will liaise with the FIA race director and attend all the meetings set up by the FIA. The sporting director needs to fully understand the sporting regulations and is responsible for bringing a driver before the race stewards if the driver has broken some rule or got himself into trouble. Because of this quite heavy responsibility on behalf of the team, the sporting director often works quite closely with strategy when trying to find a video to defend an allegation against a driver or, perhaps, recall an example of a similar infringement in the past, where the outcome back then could favour his defence of the driver or team in this instance.

And then we have the strategy group. The various teams do it slightly differently but, usually, there will be a senior strategy engineer, or the head of race strategy, on the pit wall. Like many of the functions in each team, strategy on the pit wall works with mission control in providing a lot of analysis. It is very structured. As a race strategist on the pit wall, I worked with a senior strategy engineer in the factory. There would also be two junior strategy engineers in the factory, each having very specific roles, depending on whether it was a Friday, Saturday or Sunday.

It was very clear to everyone exactly what they were covering at any point live in a session. Two people at the factory would be covering the traffic and speaking directly on the intercom to the respective race engineers, thus relieving me of that worry when working at the pit wall. I would be thinking about the bigger picture: what are we trying to do? What are we trying to achieve? Are the cars conflicting in some way? Is a decision on one side of the garage affecting the other side?

The mechanics' role in between races in the European season depended on whether the cars were coming back to the factory. Sometimes that was not the case because of back-to-back races, or the geography meant it did not make sense to haul everything all the way back to the UK and then, perhaps, have to set off again in a matter of days. In which case, some bits might be sent back for servicing.

If the cars did come back because there was a suitable gap in the schedule, the mechanics would come to the factory and strip and rebuild their individual cars, predominantly keeping the roles they had at the track by working, say, on the front or the rear of the car. Because the race mechanics are away so much, they don't appear at the factory that often. They might come in and do some pit-stop practice using a rig in the factory. Perhaps they might do some training. Mechanics would not necessarily do a lot of work on the cars between races.

During the off-season, the mechanics might find themselves building bits of kit for the garage. It could be a new stand for the team members operating from the pit wall, or perhaps some new toolboxes, or servicing some older stuff that has come back. Then, of course, they will learn about building the new car once the parts become available in the new year.

By which stage the logistics manager is finalising plans to have the cars and team kit shipped to the races. That job varies greatly between taking everything by road to the European races and preparing for the so-called flyaways to grands prix further afield.

For the races in Europe, the cars are carried in articulated trucks, accompanied by similar pop-up trucks which seem to become more elaborate each year. The upper floor rises to provide offices above what is either a workshop area or more offices. Some of the floors will include driver rooms and management offices. Quite often, the space between two trucks will be covered over to provide a tyre area.

It is a different story at the flyaways which, obviously, cannot be serviced by the trucks and are therefore less of a home-from-home. These races tend to have a pre-built hospitality centre. The engineers' office can be quite basic: chairs and fold-up tables with slots for laptops. And that is about it. The garages at flyaways look very familiar, with the usual equipment and banners in the background, but there are more freight and flight cases stacked in the paddock. Much of the garage freight travels by sea, which obviously takes longer and means there are maybe six sets rotating around the world at any time. The kit you pack after the race in Singapore might not be seen again until the final race in Abu Dhabi. It definitely will not be at the next race. I have often thought I would love to spend more time understanding how this logistical nightmare comes together, and how they cope when the late cancellation of a race – which happens occasionally – means much of the stuff ends up being in the wrong place.

Because it is not possible to have six of everything – the race cars, for example – some of the equipment needs to be shipped by air. With the cost of air freight being so high, lots of simulations on cooling specifications and wing levels are done in advance to reduce spares. The need for economy leads to the continual difficulty of only packing essentials, rather than stuff you think you might need. It is a difficult balance for everyone to achieve.

At the end of a flyaway race it is not just the packing up process that is different. From an engineering point of view, when in Europe, you do your debrief and leave the track within hours of the race finish. Sometimes, if you are lucky, there will not be enough time for the debrief, so you leave immediately. At a flyaway, however, it is usual to stay that night. This means the debriefing and downloading of everything needed for the flight home the next day is completed at the track. Then you have to clear up the office, start putting away the tables and chairs and generally try to help the guys, because everyone is trying to get everything done on the Sunday night.

The hospitality units are huge and require several trucks to carry them and a specialist crew to erect and dismantle these temporary buildings – often in a big hurry. They are your home for the best part of a week. As well as providing top-class meals, the hospitality unit is a relaxing place to be – when you have time. Quite often, particularly during the track sessions on Friday and Saturday, a quick sandwich in the garage will be the best you can manage. In a way, that is a reflection of the F1 old-school mentality of getting the job done above all else and grabbing something to eat – surviving, if you like – along the way.

I felt that was happening during my early days at Force India. Mission control was beginning to become established but the guys, because they were in the factory, did not have the luxury of the catering service we were enjoying at the race track. They were not getting dinner at night, for example. As mission control and the team grew, we pushed to change that.

At the European races, the breakfasts served in most of the hotels are not great, but the race team would only have to endure that on a Thursday morning. On the remaining days, we had breakfast at the track, as well as lunch and most dinners. That was one less thing to worry about. The quality of food is really high, even though it goes unnoticed a lot of the time. But when you stand back and think about it, you really do eat well at the track – and you need it during a busy day. It got to the point where, at the end of a European race on a Sunday, instead of being faced by a dodgy lounge meal in the airport, I would try to hang back and eat at the track.

We were looked after extremely well. I had a gripe about the difficulty when abroad of finding real milk to go with a cup of tea – an essential requirement from my point of view! This reached the point where, if I go into the Aston Martin hospitality unit today, proper milk automatically comes with my tea. You cannot ask for better than that! There was always a willingness from the catering staff to help out if, say, you were unable to eat at a certain time because you were busy and could not break away from whatever you were doing. Rather than complain, the chefs really worked hard to help out around your needs.

At one point during my time at Aston Martin, the catering became too good when they had a pastry chef in the kitchen. The desserts were amazing. Having finished the main course, you would have this notion of 'being good' and ending the meal right there. Then someone would say, 'The sticky toffee pudding is absolutely amazing! Have you tried it?' And that would be it. Dessert was too difficult to turn down.

I will never forget my last race weekend with Aston Martin, the 2022 Hungarian Grand Prix. They asked me beforehand to list some favourite foods. As a result, one of my preferred choices – duck salad, or sweet and sour chicken – was on the menu each day. It was a lovely thing to do, and typical of the chefs who, along with others such as the hospitality staff and marketing guys, go largely unnoticed and are taken for granted.

One of the earliest pieces of advice from Tom McCullough,

my manager at Force India, was to make a point of thanking everyone at the end of the race weekend. In fact, on a Thursday, I tried to find time to go in, say hello and ask how things were going. The catering crew always wanted to ask a question about the strategy in the previous race. They were really interested because, being in the background, as it were, they were not involved in briefings and did not know the full detail of what was going on.

At one of the podiums, there happened to be several new background staff on the catering team. The girls were in the garage, handing out the glasses of champagne. I told them they should come down with us to the podium because you never know how long it will be until the next one. The girls were surprised and said they did not think that was allowed. They came to the podium with the rest of the team and were thankful that someone had pushed them because, F1 being what it is, no one was going to tell them to do something like that. Podium finishes are when the race team really comes together and enjoys the moment. This is exactly what all the hard work has been about, regardless of who you are or what your role may be.

Chapter 8

WORLD TRAVELLER

It's easy to become blasé about flying round the world to the various grands prix. F1 people live in this little bubble as they travel together. Some of the experiences are amazing and you get to see some wonderful things. And yet people quickly become very negative. I found myself starting to complain about a flight being too early; or questioning why we needed to leave on a particular day when, surely, the next day would do just as well. You take for granted things that people don't normally get to do. You forget about the positives – and there are many.

When I started travelling with McLaren to grands prix in 2014, I really wanted to go racing – and enjoy it. I'd worked very hard to get into that position and I was determined to make the most of it. I remember thinking it would be brilliant to be able to say I had done a full season. I took advantage of my situation by including a lot of holidays around the races. The United States Grand Prix in Austin in 2014 was scheduled for the first weekend of November. The previous weekend, I went to Las Vegas. McLaren flew me out and I paid for my hotel. Then I flew from Nevada to Texas. Apart from giving me an amazing experience, which I might otherwise have never been able to afford, the arrangement also benefited the team because, when I arrived in Austin, I was already in the time zone and free from jet lag.

Australia also presented excellent opportunities. In 2016, with the next race being in Bahrain, I used the week in between to experience the Whitsunday Islands off the east coast. In other years, I've travelled the Great Ocean Road, or flown across Australia and visited Perth. I tried to push myself reasonably hard wherever I was. In Japan, for instance, the Suzuka circuit is beside a non-descript industrial estate. I'd go off the beaten track and try to experience the local culture by doing all sorts of stuff – like eating at a dodgy food stall on a street corner. It's a matter of not limiting your experience by confining yourself to your hotel. In Miami, the area in which we were staying and working wasn't great. But if you made the effort to go into the city, you could begin to see more of the real thing.

While some places are amazing, one or two of the hotels are really rubbish. Because we were away for longer for the so-called flyaway races outside Europe – and because, in some cases, of jet lag – the teams tended to spend good money on accommodation for those races. You stayed in comfortable hotels of a high standard with a pool, good gym and so on.

For the European races, the hotels tend not to be so nice. Even at Monaco, which is referred to as the pinnacle of motorsport, you end up in a very average hotel in Menton, just across the border in France. It's a faff to get in and out of the track, and the breakfast isn't brilliant.

Even potentially excellent hotels have their drawbacks. We had a very good one in Austria but, unfortunately, Austrian hotels don't always have air conditioning. It's often very hot, but the temptation to leave the window open leads to choosing the least-worst option: too hot, or a room full of mosquitoes?

Japanese hotels have the tiniest rooms imaginable, mainly because they're dealing with businessmen with no luggage who are staying for just one night. These rooms are not designed for travellers on the road for at least a fortnight. I don't mind, because the little rooms are always clean – which solves my main bugbear about dirty rooms. The bathroom and shower are very cramped, though you get used to it. And I love the fact that the mirror on the wall isn't quite the right height for you! It's difficult getting ready because you can't take all your stuff out of the suitcase; there's simply no room. So, yes, in some ways it's a faff, but you accept that as part and parcel of Japan. I guess the biggest problem is the little rice pillows they use in that part of the world. They're very small and uncomfortable by our standards. On one occasion with Aston Martin, team members were complaining so much that one of our drivers, Sebastian Vettel, went out and bought everyone a roll-up travel pillow. That was really good – and typical of Seb. I've kept my pillow and bring it with me each time I go to Japan.

It's all about making life as simple as possible. From a female point of view, I would have two hair dryers, one of which would live permanently in my suitcase. For the same reason, I would have two sets of makeup: one in the suitcase, because it was packed so regularly. Similarly, my back-up wash bag never left the case. This was particularly useful for the early morning departures when, if you had to pack personal items you had been using at the last minute, there was always the danger you would forget something. It was just easier to have two of the essentials.

I got good at this: I feel my long-haul pre-flight procedure is nailed down. I always have my flip-flops in my bag. The first thing I'd do on the plane would be ditch my shoes and wear my flip-flops. I always have a pair of shorts or joggers: I'd get rid of my jeans, or whatever I was wearing. It wouldn't matter whether I was planning to go to sleep or not: I would always get into something comfortable. Get my shorts on, get my flip-flops on. Occasionally I'd wear in-flight socks, but not always, because they often didn't work with the shorts and flip-flops! But that's okay; I learned not to care so much about what people thought about my appearance.

Part of my routine – particularly for long-haul flights – is ensuring I have films downloaded and a physical book in my cabin bag. I like having something I can pick up and put down really easily. And I always have my music. I don't have a pre-departure checklist, but packing so regularly means it comes automatically.

You've always got a fear of forgetting or losing your passport; that's the big concern. I've been reasonably fortunate. The only time I've forgotten something I really needed was in Bahrain one year, when it was the first race. I got there and realised I'd not packed any underwear – not a single item. We were away for two or three weeks. I still don't know how I managed to do that because I had packed socks, which were in the same drawer at home. In Bahrain, it was easy because we stayed next to a shopping centre. For some reason, I had a habit of leaving bracelets or necklaces in hotel rooms or on the side when on an aircraft. The good thing is you are travelling with this little community – and people are very helpful. Wherever you are in the world, someone will lend whatever you need.

The grand prix 'weekend' is anything but. For F1 team members, it starts when you leave home on the Tuesday or the Wednesday – depending on how far you're travelling. For the European races, the actual flight itself represents the shortest period of time during a journey that started much earlier than you might think.

I always found it frustrating to have to drive to the factory, then get on a bus for the journey to the airport. It was always much too early for my liking. Because the team's travel office tended to be risk averse, they'd have you arriving at the airport, say, two hours ahead of departure when an hour and a half would have been okay. I always felt I was in this last-minute rush to get to the factory. And then you had to haul your suitcase across the car park to the waiting coach.

Because we flew a lot with British Airways, the travelling team were members of the BA Executive Club. That meant being able to join the shorter queue for check-in and through security. Saying that, I'd find my patience being tested if the person ahead of me was completely unprepared for security and had to start fumbling in their bag or removing stuff from their pockets at the last minute. I walk quite quickly, so another of my frustrations was having someone in front of me ambling slowly through the terminal.

I think a lot of this stems from being a frequent flyer and liking to have everything organised to the point where it makes life as stress-free as possible. One of the things I found interesting about travelling with F1 team members through the airport was that everyone would head straight for the Executive Club lounge and have breakfast, or whatever meal was appropriate for the time of day. This was a ritual that was unquestioned – largely because it was free! I'd often prefer to please myself; perhaps do some shopping or buy a sandwich for the flight at somewhere like Pret a Manger, and then go to the lounge until the flight was called.

It would be the same on the return journey, particularly

if coming straight from the race track and flying back on Sunday night. The Monaco Grand Prix was probably the worst. The lounge in Nice airport is small at the best of times. On the night after a race, it would be absolutely rammed with F1 people. There was no food; it was impossible to find a seat, never mind a cup of tea. I reached the point where I'd be happy to pay for something to eat in the main concourse. I couldn't bear to be in that lounge.

By that stage, of course, your bag would have been checked in – with the hope that you would see it when the flight reached Heathrow. Losing luggage on the homeward flight was one thing; having that happen on the outbound journey could be something else. I'd hate to count the hours of my life spent in a baggage hall, studying hundreds of bags that look the same, waiting for mine to turn up. Given the number of flights I've made, I've been fortunate that my luggage has only gone missing a couple of times. On those occasions I tried very hard to remain calm.

It happened once on a flight to Bahrain. My immediate thought was that, as we were due to stay above the shopping centre, it wouldn't be a big problem. By that point, I'd done enough travelling to think, 'I can't change this. No amount of shouting at the person in the hotel reception about lost luggage is going to fix what's happened. I'm just going to run with it. Go and buy a dress, some jammies, some underwear; whatever else I need just to keep me going.' By the next morning, my bag was still missing. It was getting very close to me sitting at the pit wall in a dress! Which led to the thought: 'I can't generate team kit. If someone wants me appropriately dressed for work, they're going to have to sort something for me to wear.' And they did.

The other example of lost luggage didn't go so well – this time in Germany, where I was driving the hire van. To save time, the appointed driver would go straight to the hire car desk while someone else in the group picked up their bag. I collected the car and brought it to the front of the terminal. Everyone came out to the van with their luggage – but not mine. It hadn't turned up. They had reported the missing number from the baggage receipt I'd given them. There was nothing else that could be done at that stage.

As I was driving, I wasn't answering my phone or picking up messages. When we reached our hotel and checked my phone, there was a sequence of messages from Ryan – who had been on the next BA flight. The first text said, 'There's some of your bags on our flight carousel.' The next one – 'Oh, your bag is on our carousel!' – was accompanied by a picture of my bag, with a crying smiley face. I texted him, 'Well, did you take it?' He replied, 'Well, no. I presumed that the system knew your bag was there. So, I didn't want to take it.' I responded, 'What? So you just left it at the airport?' And he said, 'Yeah, I just left it.'

We were staying in a small town. There was nowhere to go and buy things. I ended up borrowing clothes from the other girls in the group; scrounging shampoo and whatever else I needed, just to get by. I wore my travel kit for several days because the bag didn't turn up until very late. Ryan has not lived it down! That may have been the worst example but, in fairness, considering all the travelling I've done, I've been very lucky to rarely lose a bag.

By the time you've (hopefully) picked up your bag at Heathrow on returning from a European race and got back to the factory, you're probably at your lowest ebb. Generally, as an engineer, you will have spent the previous Monday and Tuesday getting ready for the race. On the Wednesday – or sometimes the Tuesday – you travel out. You have Thursday, Friday, Saturday and Sunday at the track and then, from a European race, you return home on Sunday night. (From a flyaway, you return home on a Monday.) You've been full-on all week and you're faced with flying back on Sunday. Then, it's the bus from Heathrow to the factory. From the factory, you drive home to arrive in the middle of the night after a massively long day. Tiredness kicks in from the moment the race finishes and you leave the pit wall feeling drained of energy. You've put a lot into the race and maybe not got the result you were hoping for. You've had a few late nights leading up to the race. You feel really broken – and then you have this seven-hour journey home.

It should be said, however, that travelling with a team makes life easy in one aspect. If your flight is delayed or cancelled, or your bus doesn't turn up, someone deals with that on your behalf. Whereas now that I'm a freelance, if there is a problem of any sort, it's down to me. I must make sure that the connections are good enough; I have to double check my passport, ensure I have the correct visa and arrange transport on the other side. You don't have to do any of that when part of an F1 team.

You might find this hard to believe but, when travelling with a team, it reached the point where the only thing I needed to know about the entire trip was the time I was expected at the factory. If I achieved that, then it was not my problem any more – the bus was waiting and someone in the travel office had sorted the hotel, hire cars and everything else. After that, I was on autopilot. I wouldn't know which airport we were flying from, or the departure time.

Things have improved a lot since Covid. Before the pandemic, you went into work on the Monday. It felt like a pointless day because it seemed everyone was sitting at their computer, moving the mouse and not really getting a lot done; just turning the handle and being there for appearances' sake. I used to keep myself going with things that I knew I could easily do. The Monday mindset was to have a list of things that could be ticked off easily.

Saying that, Tom McCullough, my manager at Aston Martin, was very good. He had an 'eight hours in bed' rule. No matter when you got home, you should get at least eight hours' sleep. Since Covid, we worked Monday from home – and it made such a difference to your life. It's the simple things. You've been away for a week (particularly if it has been a flyaway) and there's no milk in the fridge. You can get your Sainsbury's order delivered. You can catch up on those bits of life that you've missed during the week away; deal with the post, get the washing done. Life instantly felt easier thanks to this change of work routine on a Monday.

When I was with Aston Martin, the washing included my team kit. It had been different at McLaren. McLaren ensured that your name was on each piece of team kit. You would return after a race and hand in your kit. Each member of the travelling team had a little locker with their name on it. Your kit, freshly laundered, would be waiting for you there. That worked well. The difference at Aston Martin was that nothing had your name on it – and there was no locker. After handing in your stuff in a bag on Monday morning, if it was back-to-back races in Europe, you would probably be leaving the office on Tuesday – and, typically, the laundry would not come back in time, which meant factoring in time on Wednesday morning to pick up your clean kit – and then packing it in your case before getting on the bus. And, you'd find, some of your stuff would not be there, or somebody had ended up with someone else's trousers or t-shirt. It was more hassle than it was worth. I got to the point of finding it easier to wash my own kit at home. Then it would be ready for packing on Tuesday night, ruling out the last-minute drama before getting on the bus on Wednesday morning – a situation that only got worse with the increase in the number of races.

The effect of the F1 calendar on your social life has not changed. The schedule for a season usually comes out halfway through the preceding year. So, you know for sure which weekends will not be free to attend weddings or family events, or whatever. I actually see that as a benefit. There are many jobs where you don't have your calendar at the beginning of the year and it's difficult to plan ahead. There's a power in knowing when you're busy.

But, of course, 23 or 24 races means you're not available for a big chunk of the year, particularly in the summer. F1 is unlike other jobs in which, if you arrange well in advance, you can ask for Friday off because, say, you need to travel some distance to a social event on the Saturday. But F1 race weekends are set in stone. No time off. F1 teams are moving to improve this.

Leaving Aston Martin and travelling to the races as part of the Sky Sports F1 team has brought quite a change to the travel process, particularly when it comes to packing. Previously, I simply brought enough team kit for however many days. If it was, say, a European race, I'd know that there would be a maximum of two nights when I might be eating out and would need to bring a couple of options for something to wear. And that would be it. Now, I must pack for each of the days and each of the evenings. I need to bear in mind that I might need something different to wear into the track. And then there's the weather. When I went to Zandvoort for the 2023 Dutch Grand Prix, it was sunny and dry when I left the UK and the forecast was for much of the same. It turned out to be wet every day – and my packing hadn't really catered for that. With a team, you brought a jacket and maybe one fleece. And that was it.

Regardless of who I'm working with, one part of the return travel routine remains the same. If I'm leaving, say, early on a Monday morning, I make sure everything is gathered together and packed before going out to dinner on the Sunday. The last thing you want to be doing at the crack of dawn on Monday is trying to make sure you've packed everything after a few beers the night before. That's another of those travel tips you learn the hard way while flying round the world . . .

Chapter 9

THE OFF-SEASON

The F1 season finishes at the end of November – and sometimes sneaks into December. That means more or less two months of no track action and more time than you have ever spent before in the factory. A misconception is that you don't do much during that period. In fact, you will be completing reports and analysis for the season just finished and getting on with planning for the year to come.

But it is also true to say that the trackside team will be using this time to claim holiday allowance that could not be taken during the busy season. One day off will be given in lieu of each weekend worked, the favourite option being to take a Friday and have a long weekend between races. However, the races outweigh the opportunities to take lieu days, particularly when there are double-headers. The other factor is that bank holidays (apart from Christmas and the New Year) have no real meaning in F1. Go into any design office or workshop on a bank holiday and it will look like a normal Monday, with everyone (who wasn't working the weekend) working flat out.

But in the off-season, before considering holidays, the strategist must get into end-of-season analysis. This will involve looking at where the car was strong or weak; where we gained or lost points as per a points tally kept at the end of every race. The totals might indicate that strategy lost 30 points during the season. If we had won the 30 points, we could have finished say, fourth instead of fifth in the Constructors' Championship. And that alone could be worth millions of dollars to the team. The strategist will go through the same analysis for other departments, building a picture of exactly what the car and the team accomplished during the season.

Race engineers are keen to know how their driver performed. You might do a qualifying head-to-head. This is not simply the totals for qualifying, or for the races themselves (like you see on the TV graphics), but a drill down to the percentage difference in lap times and the reason for that variation between drivers. The figures will be corrected if one driver did not make it through Q1 or Q2, or for any issues the car may have had, or technical updates available for one driver but not the other.

These details are important. If you overlook any of the aforementioned caveats and say one driver was worse or better than the other, it can become political! The race engineers need this analysis as they try to establish an overview and help their driver improve year-on-year. So, the first month of the off-season involves analysis of what could have been done; basically, it is a review process for us all.

Tyre performance generates a lot of interest. As a strategist, I would look at how accurate our tyre model predictions had been at each event and compare them to previous years'. Was there something that we missed and ought to be included in the model for the coming season? Some of the requests might make the strategist realise it could be useful to track the points in question during the following year by creating a report or template on, say, driver performance, qualifying statistics or tyre degradation at every race.

December is also a good moment to fine-tune tools, the past season having been so hectic that there was very little time to develop a piece of software or a spreadsheet that had an irritating niggle you had to cope with.

The off-season is also useful for sitting down with the team to generate ideas. Discussion could canvas views on what we think cost the most man hours on a Friday or a Saturday night, and how we might improve the systems accordingly. It usually boils down to seeing what might save time or create the most positive effect, and establishing which aspect will provide the biggest gain.

The strategist might think back to a wet race which prompted the question of how tyre temperature affected the cross-over from an intermediate rain tyre to a full wet – during the season there wouldn't have been time to dig into such intriguing detail. The off-season provides the perfect opportunity to carry out a full analysis and some interesting bits of research, though. The results might not add to the strategy process. On the other hand, they could come into play at a crucial moment in a future race.

Inevitably, the FIA will issue revised regulations. Now is the time to dive in, have a thorough read and mark all the changes before producing various documents on how you think the new rules might influence things. Apart from strategy, other departments could well be affected by any changes, be they to fuel, setup or Parc Fermé regulations. The strategist will ensure that the appropriate people are fully aware of the rule changes that affect them.

It can happen that the comparatively late appearance of revised regulations will affect something the strategist is working on: maybe software, a tool or, perhaps, a spreadsheet. It could be that a change to the rule governing which tyres can be used during qualifying will not work with your existing software, so the programme needs to be altered.

Such alterations also apply to drivers switching teams for the coming season or, as in our case, the team changing name from Force India to Racing Point to Aston Martin. It may seem a silly detail, but it takes time to alter the little team logos on your reports. Whatever it is, the bottom line is to ensure that everything is fully functional when you arrive at the first race. Once the season begins, there will be no time for fixing fiddly procedural details.

The winter is also a good time to update or upgrade and learn something new. I would take the opportunity to pick the brains of the bright young members of the group and improve my knowledge of software coding, say. The object-ive would be to streamline processes in order to make our Friday nights at the race track easier, simpler, and quicker.

The off-season also presents a perfect opportunity to build the confidence of junior team members by dealing with live timing information or carrying out role plays with race engineers. We would use the intercom to run through various scenarios in a race or practice as a means of increasing their understanding of what will happen during a busy race weekend. It could be that one of the younger guys had no experience of the tyre degradation analysis carried out on a Friday night. We would take time to go through that procedure and try to come up with the same answer on some previous race examples.

Health and fitness would come under the spotlight too. A gym class was always much funnier with the engineers than the mechanics, because members of the pit crew, unlike the rest of us, are seriously fit. There would, of course, be plenty of pit-stop practice, the resulting videos and reports being closely examined for means of improvement – no matter how small. The close season is an opportune time to try moving crew members to slightly different positions or modifying their technique.

The software department will produce updates which you will test and verify that everything is working as you want. Someone might have a new laptop which needs to be set up to ensure it is going to work as required when the time comes. It all boils down to ensuring that nothing will trip you up once the serious business begins on track.

The circumstances surrounding Force India over the years meant the team moved forward from being the comparatively small outfit I joined in 2015 to a much bigger operation. As a result, there was plenty of development available: it always seemed the extensive workload I had on my to-do list was outweighed by the time available to do it. It had to be a case of prioritising the items that felt the most important. Whatever the various scenarios, the target was always to start the new season in a better place than we had finished the year before by establishing what would give the most performance.

For members of the race team, the off-season seemed very different to our normal life. Going to the factory and experiencing a Monday-to-Friday, nine-to-five working-week routine soon became comparatively mundane. Apart from spending so much more time than usual at the factory, you were not accustomed to driving in and out every day. You became irritated by small things like using a lot more petrol, and the daylight hours being short in January. Finding that time of year a struggle made me want to get back racing again.

The mood began to lift, however, as the new car came together in the workshop. I would get quite heavily involved when pit-stop, or pit-wall equipment was discussed. I would be particularly interested if there were pit-stop implications in the design of, say, brakes, axles, wheels or rear crash structure. Because of my background in design and mechanical engineering, I would often take a walk through the workshop (while accepting fully that there was always a fine line between the guys being really busy and you getting in the way, versus simply keeping in touch). Watching the car come together seemed incredibly slow at first as bits were tweaked or changed. Then came a point when, suddenly, it felt as if everything changed overnight.

The first build is always the most difficult because of the number of new components. When I was heavily involved in gearbox design at McLaren, the first gearbox build was always very challenging in terms of how the pieces went together. It was very intricate. Does this bit go in front of that one? Or is it the other way round? There was usually something that clashed and refused to fit at the first attempt.

The mechanics and engineers work in shift patterns as they build the car together. There always seems to be a host of bits lying around waiting for that one piece that needs to go in first. When that piece finally arrives, everything suddenly fits in place like a 3D jigsaw. The engine and its many ancillary parts will have been installed by this stage. Now comes the 'fire-up', the moment when the engine and car come to life as one unit.

Given the mass of equipment involved – particularly the complex electronics – it's never a given that the engine is going to fire up. When it bursts into life, this is a special moment that everyone in the factory wants to be a part of. Despite the significance of the fire-up, I was always a little reluctant to join in. I don't know whether it was my background or personality, but I was aware of how busy the guys would be as they worked so hard to get the car built. Tiredness would be creeping in, and the last thing they would need is a load of engineers, many of whom they had never seen before, standing around waiting to see whether or not the car fired up.

Something on the computer might not have been inputted correctly. If fault-finding becomes necessary, they must go through the process of checking for leaks or loose connections. That's why, if I was not intricately involved in the fire-up, I would listen from the office upstairs, because you can easily hear that unmistakeable sound. You experience the feeling without necessarily being there.

Saying that, much depends on the size of the factory. McLaren was big enough to have Tensa barriers put in place to keep everyone at a reasonable distance and stop people from pushing in. Force India was too small for that, and the workshop would quickly become overcrowded with onlookers who, strictly speaking, should have been watching from a distance.

I liken it to the pit garage after an accident on track and there is hectic activity sorting everything out. I felt that my level of expertise as a strategist was not going to help in any way. Being in the garage would simply create one more person that needed to be moved aside. That, in my view, also applied to members of the management team standing around when there was a scene in the garage. Why were they there? The mechanics did not need to be told to work faster or harder.

Meanwhile, back in the factory, as the first build was taking place there would be a lot of work going on in the simulator as attempts were made to predict where the new car's performance might be. They might run the old car on the simulator to give a baseline for comparison with the new. The strategist would be keeping an eye on the predicted lap times to get a feel for where the car might be for the first pre-season test. The feeling then would be that the new season was approaching fast. Very fast.

Chapter 10

THE TRUTH
OF TESTING

The significance of winter testing depends a great deal on how the technical regulations have changed for the new season. If there has been a big swing, the cars appear for the first time looking very different. You may have got an impression from the launch of a rival's car although, in some cases, teams give very little away by largely taking the previous year's car and painting it in a new livery to make the launch worthwhile from a marketing point of view. Or they could show the latest car with bits on it that, in truth, would never work in real life.

The launch is a massive distraction in the factory. The last thing the various departments need is to be designing and producing what are, in effect, rapid prototype parts just to make the car look good. You can sense the frustration over having to push a launch car out the door when, in truth, everyone wants to crack on and get the race car completed in the limited time available.

There was a tendency a few years ago to do away with a separate launch in favour of simply rolling the car out for everyone to see at the beginning of testing. More recently, launches have come back into favour and sometimes coincide with so-called 'filming days' limited to 200 kilometres and purportedly for promotional purposes.

Given that winter testing is now limited to three or four days, that 'filming day' is very useful, particularly from the mechanical side, because you could lose half a day in testing if the car has some problem or other. It is amazing to think that one week can make quite a difference to the aero development and the build programme. It means there must be a compromise between getting the car ready for the first test and having a shakedown in advance of the test.

As a performance engineer at McLaren, I was very heavily involved in testing. Apart from finally seeing the complete car, testing provides an ideal opportunity to sample many things: from new systems and procedures, to the simple process of learning how the closing of brake-duct air flow differs on the new car compared to the previous one.

While you are learning, the mechanics are becoming accustomed to building the car. It means working very closely with this group of people in finishing the car, taking it to track and getting the car as close as possible to the right specification – which is limited by, say, only two of the ten damper springs being available at this early stage.

For the first test, it is often a case of the specification being set for you, rather than the other way around.

As a strategist, I approached testing differently. In the days when there was more than one pre-season test, the information coming from the first test was very limited. Unlike today, detail of the tyre compounds being used by the various teams was not widely available. That meant stationing a team member at the pit exit and noting the tyre colours being used. This information would allow the strategist to begin building a picture of what everyone else was up to and how the various tyres were performing.

That said, there was not enough information emerging from the first test to warrant a strategist's presence. It is not like a race meeting where the strategist is analysing data for the next practice session. The focus at the test is raising the car's operation onto the right level. As a result, I rarely attended a test session. Much as I might have had a very strong desire to see the car run for the first time, particularly if there had been a major change to the regulations, I did not feel the need to spend 18 hours of my day trackside.

In any case, the working environment for a strategist is very different at a test session compared to a race weekend. On the Friday of a grand prix, there are two one-hour sessions and we would spend most of the evening analysing the data. A test session means an eight-hour day of track action. This produces a massive volume of information that, with the best will in the world, cannot be analysed as quickly as we would like because the data simply outweighs the available hours. The engineering team would find themselves doing crazy hours at the track.

Also, remember that, at any test, there are only ten cars running; one per team, as mandated by the regulations. There is none of the usual concern about avoiding traffic. It explains why the running joke for a team during testing is to sum it up by simply saying, 'Fuel, tyres, send.' A strategist is not needed for that! But I did feel guilty at times, particularly when there would be the test at Bahrain and then straight into the first race at the same circuit. The guys on the race team would have already been there for almost two weeks when I rocked up, bright and fresh, at the end of week two.

One of the benefits of not being at the track during testing was using the time as a rare opportunity to work in mission control with the other strategists. It was also a very useful period for teaching and monitoring the junior team members. The same would apply to the mid-season tyre test – usually held immediately after the Hungarian Grand Prix in July. This test tends to be limited in terms of the tyres being used which, in any case, have little or no bearing on the race that has just finished and the one coming up next – all of which the strategist is fully focused on.

For all these tests – pre-season and summer – we would work shifts in mission control; one strategist covering the morning, before someone else took over for the rest of the day. The aim was to spread things out and avoid working people too hard during that initial stage of the season – which could prove difficult because everyone was keen to get involved with a new car and it was sometimes necessary to put out a reminder that the season was going to be very long! When the tests began in Bahrain, for example, it meant either an early start or a slightly off-set day at mission control. I never felt it was worth overdoing it at this stage of the season. There would be time for that later!

When it came to testing, 'pre-season keenness' was something you had to deal with. The new car would be running and, sitting in mission control, you would have only a limited amount of feedback and information at that stage. Nonetheless, I could guarantee that, towards the end of the day, someone from management would come into mission control and ask your opinion on where the pace of the car might be compared to everyone else. There was no easy answer to that. You didn't know the fuel loads being run, or the engine modes. And, as I mentioned above, in the early days there was no knowledge of the tyre compounds being employed. Making a bold prediction on pace was almost impossible because so much was unknown.

Teams approach testing in different ways. Some do lots of low-fuel runs while others prefer not to carry less than 30 or 40 kilograms of fuel. On one occasion, Mercedes chose not to run the softer tyre compounds at any stage, which meant it was very difficult to get a read on their potential. Added to which, the track condition varies massively from one day to the next, with lap times also being affected by the way in which the track rubbers in and gets faster throughout the session. There is the continuous conundrum of the track improving in one way and yet becoming slower as the temperature increases. The strategist dealing with these opposing elements affecting the track – not to mention the wind direction possibly changing throughout the day – finds it difficult to compare one day with another.

Surprising as it may seem, assessing drivers in the same team is also difficult. The team will run extra sensors in the car for the first two days before they are removed. This changes the weight of the car, as does, of course, the amount of fuel on board, all of which affects lap time and could possibly lead to wrong conclusions when one driver appears to be slower than the other.

Engines might be running at particular levels of power; there is no requirement to run at the legal weight during a test; the amount of ERS (battery deployment) in use can vary, as can the way the driver may or may not be pushing on the straight. Is the car running with or without DRS (drag reduction system)? There was a year when Red Bull stopped rivals from estimating their straight-line speed by continually turning off the DRS before the braking area. The strategist is faced with interpreting all these facts when someone asks, 'How are we compared to everyone else?'

While speed over a single lap is difficult to judge, getting a fix on race pace is easier. One important target – particularly for the mechanical side – is to run a full race distance. When testing at Barcelona, the best way to establish pace for the 66-lap race was to do three 22-lap stints. That allowed long runs, possibly on three different tyre compounds. This also allowed a race trace: a plot of how the race might run, particularly if you can correlate the pace of various teams at roughly the same time of day, even though that might mean making corrections across different days and different tyres. You can see how it quickly becomes quite complex.

A 22-lap stint allows you to take out a few unknowns because the calculation is based on starting with 100 kilograms of fuel and finishing on close to zero. I was always interested in the long runs, which tended to provide more reliable information than one-lap pace, which had so many unknowns. Taking into account fuel effect, a lap with traffic or, perhaps, a mistake by the driver, it is feasible to work out an average lap time across a long run. With this level of filtering and interpolation, it is possible to have more accurate data than the information provided by a single lap.

Race trace is a graphical way used by strategists to represent lap time across various drivers. We plot each lap time using a reference, which would be a horizontal line. If the driver did the same lap time every lap, the graph would simply be a straight line. In reality, the graph will rise and fall depending on faster laps or a tyre stop which calls for slow laps in and out of the pits.

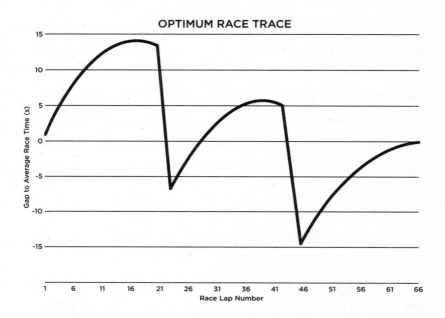

Despite this see-saw effect, the overall picture shows a slower driver gradually dropping down the graph when compared with the fastest driver, who should start and finish at zero. Most teams display race pace in this way, with the fastest driver at the top and the slower drivers below. Every team uses more or less identical graphical representation because people move across teams so it makes sense for the same software to be in common use. It is a handy way of seeing where the pit stops are, how the lap time evolves.

OPTIMUM RACE TRACE

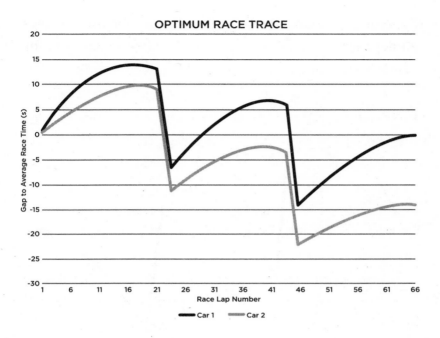

But the emphasis may be changing. With fewer and fewer test days, and cars becoming more and more reliable, there is less urgency attached to completing a full race distance – which will suit the teams because it means less opportunity for rival strategists to calculate their race pace. This will make life even more difficult for the strategist when dealing with questions about the pace of their own car. Even throwing in the caveats mentioned previously, people still want a definitive answer.

Another factor unique to testing is having just one car running throughout. With everyone working on the single chassis, focus can begin to wane as the days go by. Normally, with two cars, each team member is regimented and working solidly on their specific role. With the single car in testing, there is time for everyone to have an opinion about issues in which they might not otherwise be involved. People become more excitable and then, towards the end of the week, tiredness creeps in and adds further to the rather hectic atmosphere. Throw in an opinion from mission control that has not gone down well because the on-track team have had a full-on day, and you begin to understand why pre-season testing can be stressful – even for strategy.

All of which does not end with testing. Predictions for qualifying or race performances at the first grand prix are difficult because you continue to have minimal information. It is possible to use data from the previous year, with the proviso that every team will have made a leap in performance in the 12 months since the last visit to this track. Obviously, some elements of winter testing can be included but a solid marker on performance will not emerge until at least qualifying and a single race have been completed.

While all of this is going on, the strategist is getting their head around administrative technicalities connected with the various databases and graphs. Each driver and team is represented by different colours. When a driver moves team, it takes ages to associate that driver with a different colour. Similarly, a team's representation could change because of their new sponsorship and associated colour scheme. For example, Racing Point was pink for many years but, when the BWT sponsor switched to Alpine, it was easy in the heat of the moment to forget that the colour pink had also changed on your graph.

We would try to follow the same order favoured by the FIA, which is based on the teams' championship finishing positions from the previous year, with the driver order being within that. By having our graphs and tables in the same sequence it became very easy to find the team or driver you were looking for. The downside was that muscle memory was difficult to shift when going into a new season and colours and positions had changed.

We would do everything we could to make information easy to read or identify. Within each team, there would be Driver 1 and Driver 2, using the same order as established by the team and demonstrated by the camera colour on each car's roll hoop. At Mercedes in 2022, for example, George Russell had the black Driver 1 and Lewis Hamilton was Driver 2 in our terms because he had the yellow camera. We would distinguish between drivers on our graphs by the Driver 1 showing in squares, with the Driver 2 in dots; similarly, one could also be a solid line, the other a dashed line. For Hamilton and Russell, these lines would be in turquoise because that, for our purposes, was the Mercedes colour. The Mercedes drivers switched a few years ago when Lewis wanted the yellow camera to match his personal choice. That took a bit of getting used to and it would really annoy me if I issued a report showing the drivers with the wrong designations or colours!

Occasionally, when a new driver with an associated new acronym came along, nobody would know who you were talking about or, perhaps, how to pronounce his name. I was terrible when it came to that. For some reason, I could never say 'Vandoorne' (as in Stoffel Vandoorne, McLaren's driver from 2016 until 2018). I would give my version of 'Vandoorne' and that would be religiously repeated – wrongly – all the way down the line.

Testing is a good opportunity for the whole team to get back into the swing of things. The danger lies in feeling you are getting on top of it while forgetting there are only ten cars running. Move on to the first race and suddenly it is double the number of cars and radio traffic.

Typical of F1 lateral thinking, you learn to make good use of every opportunity, particularly if the first race is at the same venue used for testing. We would set up both sides of the garage, running the car out of one side on one day and swapping to the opposite side for the following day. That way, the infrastructure could be checked out in preparation for having two cars in the garage on race weekend.

With a reduction in recent years of the number of days of testing, some teams have found it preferable to switch drivers during the day – which can be another cause of confusion. Or, perhaps, run the first chassis at the opening test and bring chassis number 2 to the second test. Although that allows both cars to be checked out in advance, not every team has the necessary resources to have two chassis ready for testing.

Generally, we would operate from one side of the garage or the other because it would be too complicated to switch midway through the day. But that in itself can bring potential problems. It means both drivers, during the test, are operating from the same side of the garage when, in reality, one of them will be on the opposite side at the start of the season. That causes confusion because the buttons with drivers' names on the intercom panel are not correct for one of them during testing and you can guarantee someone will use the wrong button at some point. On top of that, the pit-wall layout may have changed, or the infrastructure in the garage is not the same as it was when the previous season ended in Abu Dhabi. All this needs to be sorted out, not only physically, but in everyone's head by the time everything goes full-on at the first race.

When it comes to what is happening on the track – which, after all, is the main purpose of testing – it is very easy to become confused about which tests are important. There are always more items in need of checking than time allows. As an example, take trying a different front wing. In an ideal world, you would run both wings with tyres of a similar age and in near-identical track conditions, while keeping the same fuel level and not changing any settings. Because track time is too valuable for that, something else will need changing on the car, in which case every effort is made to ensure these changes will have minimal effect. While that may be reasonably within your control, there is nothing that can be done about a variation in track condition or temperature, or the amount of traffic on track. It is very, very difficult to get pure back-to-back tests, which means there is always a necessary element of management.

The biggest difficulty in all of this is having everyone on the same page. Returning to the front-wing test: you could plan two comparison runs and, in the meantime, someone decides to try something different with the gearshifts, in the mistaken belief that it will not have an effect. But it does: the wing test has been made useless because someone has not been listening to how important it is not to change anything, no matter how seemingly minor. Similarly, if the engine is running too hot or too cold, increasing the air flow – or reducing it – on the cooling ducts may seem a simple operation, but it will change the car's aerodynamics. It is important to keep everybody listening and paying attention at all times – which can be a challenge during an eight-hour day.

The problem is that the various departments are understandably focused on their own interests. Aero will want to test the wings and the ride heights; they will not care about the car oversteering or understeering. The driver will need to check the balance of the car and find out if it is easy to drive. Then he will want to run low fuel and the softest tyres to find out what he needs to go faster. The race engineer is looking for the ideal setup, and the tyre guy wants to have a straight comparison between compounds. Overarching all of this is the number one goal, which is achieving reliability during testing.

Going back to the front-wing example: it is an important test, but the power-unit people might say that the engine will not be able to cope with running that hot again. This is where the communications can become difficult. You may have a very good and valid point. If that outweighs other sound and rational points it becomes a question of knowing when it is important to speak up. There is simply not enough time to debate everything before each run. It becomes a case of deciding whether to simply note a concern for later discussion, or act now because one particular issue threatens to stop the car.

At such times it is very useful to be working with people who have 'been there and done it'; personnel with strong trackside knowledge and a broad understanding of how the various groups work and what their concerns might be. It is collective experience that gets you through this complex process.

Meanwhile, the strategist's role is to search as best as he or she can through the mass of information, load up anything that has not been corrupted in some way, look for standout results and, if they do not ring true, try to discover the reason why.

Despite what can be a minefield of results and information to sift through, the ultimate aim, in basic terms, is ascertaining how the driver feels about the car. As mentioned, we cannot rely strictly on lap time because we don't know how it stacks up compared to everyone else's. But the driver can tell quite quickly how predictable the car is and how it reacts to the changes made to the setup or, say, the front wings. At the end of the test, if the driver is very unhappy, the chances are you are in for a difficult year. That can be very hard to take at such an early stage of the season.

Chapter 11

REVEALING YOUR HAND

In recent years, Bahrain holding pre-season tests and staging the first grand prix has brought to light both advantages and disadvantages. The most obvious plus, when testing has finished, is not needing to strip the garage, ship it somewhere else and start all over again. As pointed out in the previous chapter, the infrastructure will have been tested and proven, ready to run for the first race. With a bit of luck, some crew members might enjoy a rare day off in between, but the chances are slim as they will be involved in building a second car while on site. The engineers will be working their way through a pile of analysis in preparation for the race.

A downside comes when the need to fix a major car fault means getting parts quickly to Bahrain. When testing at Barcelona, a new component could be machined in the factory and driven to Spain in time for the next day's test. A quick turnaround for Bahrain is complicated by the need for air freight and a more involved customs process.

On the other hand, Bahrain presents hotter and more reliable weather conditions than Spain, where it can be cold at times, with the chance of a morning or afternoon lost to rain. It is possible to run in the wet, of course, and check out reliability issues and rain tyres, but, overall, it is frustrating, given the limited number of days allowed for testing. Little will be learned in the rain about aero, and the truth is that only a very small proportion of the season ahead will be run in similar weather. There is also the risk of your driver going off in the wet and, say, losing a front wing. Because this is such an early stage in the production process at the factory, the chances are that this wing, or other part, is the only one available – which is going to make life very difficult, cost more valuable time and often means fitting a less preferable part.

In Bahrain it is a nice feeling to go to the track on a warm, sunny morning and know you are guaranteed a full day's running in dry conditions. Drivers have been known to crash in the dry, of course, but even if it is not a big shunt, a major problem can result during testing. It could be something relatively simple, such as running a kerb aggressively and damaging a brake duct or, much worse, part of the floor. Even if the breakage is limited to an edge or a corner of the floor, because the floor is so vital to the running of the car, it will seriously affect the testing of so many factors. With the best will in the world, assuming repairs can be carried out in the garage, the floor will not be the same quality.

Ideally, of course, you do not want the driver to be running a different line, never mind using the kerb aggressively. Consistency during testing is paramount, otherwise lap times are not representative. It is all about getting a driver into the headspace of producing identical laps, one after the other. Contrary to what is expected on a race weekend, the driver should not be looking to constantly improve lap time. It should be about each lap having the same line, similar levels of pushing and comparable feedback.

This needs to be a pure test of a component, tyre, setup change, or whatever is being checked out. You do not want the driver to say, 'Before, I was only trying for 50 per cent of the lap. This time, I pushed flat out, so that's where the lap time is.' Having the same inputs again and again to get a consistent reading may be what you want, but is actually very difficult to achieve.

Development is ongoing from the moment the test starts. If, say, testing comprises two groups of three days, one week apart, then an upgrade might arrive for the very last day of testing because, during the week and a half since testing started, a development worth running has been found by aero design in the wind tunnel. The upgrade that arrives for the last day of testing is normally the package for the first race. In broad terms, there can be a launch package, a test package and then a Race 1 update. But, overall, the search for improvement is non-stop, with the last day of testing representing the latest thinking.

The drivers will know this and there will be a friendly fight to see who gets behind the wheel on the last day of the test when, in theory, the car should be in its most competitive state thus far. All the aero tests will have been completed and, hopefully, early niggles sorted out. In addition, many of the sensors will have been removed, which could reduce the car's weight by as much as 12 kilograms. This would include taking out extra thermocouples to assess the temperatures of the engine or the gearbox; additional sensors for all sorts of things such as monitoring fluid pick up when cornering or double-checking fluid levels that, under normal circumstances, would be assumed to be okay. Less data will be produced, but the car in this state begins to hint at what might be possible when the first race weekend gets under way.

The final day of testing is also used to practise live pit stops. Up to this point the crew might have been working shifts, but everyone will be present in the garage for the final afternoon. During the previous few days, the drivers will have been rehearsing stopping on their marks in the pit box at the end of each run as part of the acclimatisation process with the new car and brakes. The drivers might also try launching out of the box, as if in a pit stop.

Pit-stop practice is also an opportunity for new pit-crew members – or someone who has changed position – to feel comfortable with the drill. The crew will take time out to practise turning the car around quickly in the garage during qualifying, the target being 45 seconds to have the car come to a halt, refuel and get going again.

Procedures will increase if the driver is new to the team or, even more importantly, a novice about to start his first season of F1. Steering wheel and pedal setup will have already been done but testing gives the opportunity to run through the steering wheel's switch positions and experience the effect they have on the car in real time. There will be lots of defaults which, in simple terms, are a reset similar to turning your computer off and on again. But they are often very complicated to do on the steering wheel and this is an ideal opportunity for the new driver to become accustomed to the systems rather than in the heat of the moment during a grand prix weekend.

At some stage during testing, usually on the final day, the teams will do a fuel run-out check, which is exactly as it says. The driver keeps going until the last of the fuel has been picked up and the car coasts to a halt. The accompanying high number of red flags will screw up your run programme if you are not prepared for cars stopping on track at this late stage of the test.

The fuel run-out is part of a long list of procedures that need to be checked off. It varies from team to team but prioritising this list – deciding what is done and when – is usually down to the head of engineering. The list will be discussed during briefings in advance of the test, or at the start of each day, depending on what has happened in the previous 24 hours.

There will be requests from the power-unit supplier (in the case of my time with Force India through to Aston Martin, it was Mercedes) to check out items such as selecting and running the car in reverse. They will also want to try the anti-stall and, at some stage, carry out a heat soak test. This involves running the engine quite hot and leaving it without the application of fans to see how quickly it will overheat. The data will be monitored to establish the critical point at which fans will need to be applied to avoid the risk of serious damage being done. All of this will be part of a package of tests the engine supplier needs to complete before signing off the power unit in readiness for the season ahead. It is why one senior team member in engineering needs to be responsible for bringing all the various departments together to ensure the complex matrix of checks is carried out.

Because there is so much ground to cover, things can become a bit tense as time begins to run out towards the end of the test. The power-unit engineer might suddenly say, 'Oh! We've missed this! It needs to be done.' Or the gearbox guy might speak up and say, 'Look, this is really important; we need to get a read on it before the end of the test.'

The daily plan continually needs adjusting because runs have been interrupted by a technical problem or a red flag.

Therefore, the senior member must think on his feet and reorder priorities. A red flag, for instance, could make it obvious that the track will be closed for at least 30 minutes. Rather than sit idly by and waste time, the on-track lull could be an opportunity to complete a time-costly setup change. Decisions also need to be made if, say, a request to test reverse brings the question, 'Yes, but what if it fails and we can't run for the rest of the day because the car has been damaged?' This is where experience is needed to ensure the important work is completed. Yet again, good communication is essential.

While all of this is going on, the team will have a photographer working trackside and in the pit lane. A lot can be learned from images of rival cars. It is common to see photographers staying in one spot at the end of the pit lane and taking images of each car from exactly the same angle. A comparison between these and your own car allows engineers to use a mapping process to gauge the ride height or wheelbase being run by rivals. Or you might notice that another team has introduced a new front or rear jack and it would be useful to have a photo of it to see if there is anything clever that is worth knowing.

It is all about building a base of knowledge rather than simply copying a ride height or making a replica of, say, someone's front wing. A straight copy simply would not work on your car because so much research has gone into the very fine detail. But, then again, variations of different ideas can be tried in the wind tunnel on the basis that something, however small and seemingly insignificant, might be learned.

A quick look at a rival team's pit wall might also reveal valuable insight into the design, be it the layout or the on-screen graphics which can be seen from a distance. When it comes to pit-stop equipment, anything goes. Ferrari was one of the first teams to move away from a man with a lollipop to lights signalling when the driver is good to go. Now everyone has lights. The pit-stop process converges along the pit lane thanks to teams watching each other and analysing every detail in the search for a tenth of a second saved.

When working in and around the garage, the golden rule is to remember that there are eyes everywhere. It is more than your life's worth to walk through the pit lane with an exposed setup sheet showing fuel load or ride height figures. Photographers are constantly at work, which explains why many team personnel can be seen walking around with their work and information sheets folded in on themselves.

Given half a chance, photographers will take images of computer screens in the garage. If working on the pit wall, it is vital not to step away and leave a data page showing. That would be public knowledge before you have time to say, 'Log Out'. If I had been given a close rival's fuel load, that would have been one less important element to worry about. It therefore became a habit to leave the pit-wall screen showing a generic page such as lap times or details that are common knowledge.

We would try very hard not to have printouts of component assemblies and such like because it would obviously be bad news should they go missing. Things change so rapidly that drawings can quickly become redundant if, say, a bolt does not fit as it should. That means a new part and associated assembly drawing, with the old diagram no longer needed but still providing useful information if it falls into the wrong hands. During a season, you become aware of which photographer works for which team. As the last person to leave the pit wall, I would ensure everyone's screen was locked even though I knew someone from IT would, at some stage, come across from the garage and check everything was shut down.

The worst-case snooping scenario would occur during a grand prix in the period following a red flag, with the cars queuing in the pit lane, in race order. It could be, for instance, that a Ferrari had stopped beside your position on the pit wall. You would be busy on screen working out the shakeup to strategy caused by the stoppage, while being aware of a bunch of Ferrari mechanics standing directly behind you, looking at your screen – and there'd be nothing you could do about it. It is the same for everyone, of course. But slightly irritating nonetheless.

The period before the start could be just as tricky. I would stay on the pit wall while the grid procedure was taking place. Given half a chance, rival team members passing to and fro would take the opportunity to look at my screen. The trick was not to have your strategy showing at that point – but to remember to put it back up at the right moment before the really serious business got under way!

Chapter 12

THINKING AHEAD

No sooner has one grand prix finished than the strategist will be thinking about whether any of the lessons learned can be carried forward to the next one. It is a process that starts well before each race weekend.

If we take an event in Europe as an example, a strategist will have spent the previous Monday and Tuesday doing prep in the factory. There will be a pre-event meeting – usually on the Tuesday – when all the big players come together. The race engineers will lay out what needs to be done from their point of view, the aerodynamicists will have their thoughts, and the build department will say what pieces are available for the car or, perhaps, missing. So, you begin to formulate more of a plan for what lies ahead. In advance of that, the strategist will have produced a preview that involved a reasonable amount of work – an important document which will form the basis of everything you do for the rest of the weekend.

You will have worked out the percentage chance of a safety car during the forthcoming grand prix by checking how many were deployed in previous years, and noting when they occurred during the races. It will also be time to check the early forecasts and look at the weather patterns during the past four or five years. Overall, you ask: how did these races go? What was important? You examine the notable errors and the correct decisions made each time. The strategist is basically building a picture. You are trying to align people and channel their thoughts about what to expect. Before the Singapore Grand Prix, for example, you would remind everyone that overtaking is difficult. Or underline the point that the high probability of a safety car will mean perhaps not going for what appears to be the optimum number of laps for whatever the tyre choice might be. At this stage, you're trying to get people on the same page while mooting, early on, how their preferred choices might not be possible. The idea is to end up with a tyre model; a strategy model; a feel for what you think your pace is going to be for that weekend, and how it will compare with rivals'; a prediction of your likely qualifying and finishing positions – plus an indication of when the best time to make a pit stop (or stops) might be. Even at this early stage, you will have a simulation that tells you which strategy you think is correct.

All of this is based on historical data for the event: how the tyres have reacted in the past; was the degradation better or worse than anticipated? A trend, year-on-year, means you will end up with a matrix of available information.

During pre-event preparation, this is part of justifying your model, of making sure your assumptions and calculations are correct. Even before practice starts, you will have a rough plan of how the race will go, what the big items are and what you need to learn on a Friday. This is where the historic data is useful for another reason. If, say, you learn nothing during Friday practice because it's wet or the car hasn't run because of a technical problem, you have a model that you are reasonably comfortable to run for the rest of the weekend and which enables you to start your work.

Any rain during the sessions should have been predicted before the start of practice, if not during the early preparation. The strategist will have gone through every type of forecast, from wind direction to ambient temperature which, if nothing else, can help you decide what to pack before leaving home! Weather predictions will obviously have a big effect on tyre choice.

There are three tyre compounds: hard, medium and soft (I will describe the regulatory allocation in a later chapter). At this point, you might be saying, 'A major uncertainty in our model is the hard compound (or the medium, or the soft, depending on circumstances). If we learn just one thing on Friday, it should be about this.' You know, based on your calculations, which compound you've got less running on, or which one you've got less information on, and which tyre is most important for the race.

At the factory pre-event meeting, you're trying to say

this is what's important to learn from a strategy point of view. You end up with everyone vying for what they think is important to learn. From this mix of differing opinions, the engineers must formulate a plan for what happens on a Friday.

You use spreadsheets or databases when tracking and recording the tyres people have used in the past, checking your tyre models during that weekend, or noting the number of red flags. These databases are set up in readiness for first practice on Friday morning as part of a considerable amount of prep completed even before leaving the factory.

The data from the previous year's race would have been checked and, if time allowed, we might run a replay of a practice session to see what the traffic was like, or review a qualifying session to see how hard it was to get a free lap. Or, perhaps, check the race itself to understand how the pit-stop decisions worked out. Looking through the FIA documents, I might note that a driver got called up for crossing the white line on pit entry, prompting the question, 'Is there any way we can avoid that and other potential problems?' The software can rerun the data and you can make decisions as if the race is live. More than half of the teams use 'RaceWatch', which is a piece of software created by SBG, a company started by James Vowles, formerly the strategist with Mercedes F1 and now team principal at Williams.

The intellectual property in strategy is actually the model you put in. It could be how you predict the hard tyre is going to perform, or how you predict the car's pace. You are modelling what you think the pace of the ten cars will be and how you think the three tyres will perform; these bits are the intelligent part of your prediction for the race weekend.

Some teams can do a full interactive replay on the TV and back at mission control. They can also include some of the intercom chat that was going on at the time. Fully replaying a previous race in this way makes it very interactive. If you're trying to teach a junior member about something such as traffic at the Red Bull Ring in Austria, this could show, for instance, how quickly you need to tell a race engineer to advise their driver in order to avoid a penalty for blocking a fast-approaching car.

Minutes of the meetings will be kept. My strategy pre-event document was something everyone could read at their leisure. Other groups would do something similar. Race engineering might produce a document about the car setup plan, or the tyre guys could record their thinking at this stage. Very often there would be a disagreement between strategy and tyres over which was the best tyre. Or what each side thought was the best strategy – and why. The most common discussion is usually about whether a hard tyre will do better at the beginning or at the end of the race, based on variables such as fuel load and track temperature. The general strategy preference is to have the short stint first and the long stint at the end, with the fastest race to the final pit stop, as this sets track position. The tyre engineer, however, may strongly believe the hard tyre will be best at the race start so will be pushing to make a long first stint possible. Much of this can be data based, but it is possible to model these opinions in and get a calculated answer.

There are quite a few things that are not easy to model. Sometimes the tyre engineers won't necessarily have a justifiable or quantifiable answer for how much better the hard would be on, say, a two-degree hotter track – or whatever it is that you're trying to mix in this calculated and estimated world.

A lot of what you do is estimation. Much of it comes from what you know historically about the track and the race, and how much, say, the track has improved year on year. You know the fuel load; that's easy to calculate. And you may have an idea when the safety cars might appear, but there is an awareness that these things might not happen exactly the same this time round. Quite often, the race organisers may have resurfaced the track – or parts of it. That's an issue because, even though you have done all your pre-event work, you never really know the effect of the resurfacing until you get there.

Pirelli measure the track roughness and send that information to everyone. But you're wise to stay sceptical. Each team measures the track with more or less the same equipment. The bottom line is that you're trying to get an indication of the roughness by, in simple terms, measuring the size of the gaps between stones on the surface. Teams will usually do that at each corner. Some tracks can become cut up quite badly in certain corners, which can damage tyres.

In the end, it's a balance: you produce one number for the whole track and that goes into your model. It could mean that model is incorrect but, at this point in your pre-event prep, it is your best guess of how it will go. You are trying to decide what tyres to run on Friday, and in which order. Strategy will have one answer – but be assured there will be other points of view!

There can be unique variables for a particular track. A good example was Singapore in 2023. There were quite significant changes to the track layout, which fundamentally affected tyre performance. The strategist would have flagged that during the pre-event meetings and tried to get the team on board.

The major track alterations are likely to be documented by the FIA, but some of the smaller changes are also worth knowing. I would bring myself up to date by checking out other racing series at the same track. MotoGP, for example, also use Circuit of the Americas in Texas. I discovered that the riders had been complaining about bumps, which led to the track being resurfaced. That sort of information might not be available in any of the official documents. I would follow each circuit as best I could on Twitter, read updates which could, for example, show images of new kerbing. It's all about keeping on top of every piece of information available.

You rely much more on historical information than data from the simulator – which would have inaccuracies in any case (which we have discussed in the previous chapter on simulators). Drivers perform differently in the simulator.

For Singapore, you would look at the previous grand prix at Monza – and a few races before that – and formulate an estimated pace to give a realistic representation of where you might start, and the strategy factors that might therefore affect your race. It's important because – to take an extreme – you would potentially have a very different race from pole position than you would when starting from the back of the grid.

For a new track, you will ask the drivers to do a run through the pit lane on the simulator to help with early estimates for pit loss. That is reasonably difficult to do because you will not have accurate figures for the braking necessary at the pit-lane entrance, and also the speed on exit. You won't know at this stage in the simulation if the pit-lane speed limit lines are in the right place. It could also be that, between now and the start of practice, barriers could be moved. It is a good example of why, in a pre-event meeting, the strategist would push hard to establish the pit loss because it's something you are really worried about in your model, particularly for a new track.

Our first reference in such a case is a drawing of the circuit. It's 2D – in the same way a plan shows a house that has yet to be built. You know where the corners are, relative to each other. Then you often receive what is known as a LiDAR (Light Detection and Ranging) scan created by a car of some description physically going round the track and scanning in 3D.

At the same time, the simulator engineers will use something along the lines of Google Maps to help create surrounding images. It is very dependent, of course, on how much in advance the track is finished. Quite often it's last minute, which means there may be more resurfacing, or some kerbs and bollards yet to be installed. The kerbs are important because, depending on their size and whether or not the drivers can use them on the entry and exit of a corner, they can have a considerable effect on lap time. The cambers of the corners should not present a problem but a bump, serious enough to have the driver avoid it, could also affect your predictions.

With all this information to hand, the pre-event discussions can then focus, say, on whether it's going to be a one-stop race, which is going to be difficult and require a certain amount of management. Or, perhaps, it's going to be a two-stop and we need to save a hard or medium tyre from Friday. You want whatever prediction you've come up with to be in everyone's mind in advance of getting to the track on Thursday. You want them to know what you're trying to target – and why. The strategist has to be reasonably forceful in those discussions. In fact, it begins at the post-event meeting from the race before, when you might have highlighted things like: 'We didn't get a good read on pit loss.' It feels at times as though you are hammering home negatives to help the cause for the following race.

Some people really take it in – and others don't. The strategist must accept they are not going to get everything. So you ask yourself, 'What do we really, really need? Do we need pit loss? Or do we have loads of information for the coming race, so establishing pit loss is not a major issue? Do we need them to lay rubber in the box in FP3; is that really important from a strategy point of view?' You need to pick your fight.

Part of the strategy role is being pragmatic. Everyone wants to win the race – but, for most, that's probably not realistic. The strategist must be straight and say, 'I think we've got a chance of P10 or P11.' That's not what people want to hear, of course. But you must draw this line and be very clear. The pre-event work may say that you expect to struggle to get into the third part of qualifying; Q3. That being the case, should you be saving all your soft tyres for Q3 in the usual way when, realistically, you're not going to get that far? It is a case of setting the scene as reasonably as you can, but without being too pessimistic. Drawing that line can be difficult but it's essential if you want everyone to understand that, say, getting out of Q1 is going to be a challenge, so maybe we're not going to need all the soft tyres. Williams did that for many years: running three sets of soft tyres in FP3, because of the acceptance they were unlikely to need them, given getting into Q2 or Q3 was a big ask. So, make good use of the soft tyre in free practice.

The strategist should always be the one arguing the more sensible approach, even if it is uncomfortable to hear. On the rare occasions when your message is not being accepted, you can resort to speaking to the chief race engineer, or chief trackside engineer – it varies from team to team – essentially, someone who's above both the strategy and the race engineer. Generally, it's the race engineer who will set the run plan for the car. It is definitely the race engineer in communication with his driver.

For something you really need to happen, the race engineer – and driver – must have bought into it. Say you believe the correct strategy is to have a long run on a medium or a hard tyre. If the driver isn't in agreement with that, he's going to drive incorrectly. He will push too hard, which is not how a one-stop strategy will work and, in any case, by driving this way he will produce a result that shows high degradation. You must make it absolutely clear that there's no point in driving like that. A two-stop is not in the plan – despite what the driver might want!

The conclusions drawn from advance meetings allow time for the engineers to filter that information to the driver before he arrives at the track. By getting the bulk of preparation work done by Tuesday afternoon, before flying to the venue on Wednesday, the strategist knows that the team is becoming aligned and run plans are falling into place.

The crew at the track will have been able to crack on with setting up the car in readiness for first practice, when we begin to get some solid answers after hours of estimation and speculation.

Chapter 13

FINALLY - A TRACK!

Working at the flyaways is never as comfortable as the races in Europe. Either the seating positions in the temporary office are poor, or the aircon is not working properly – it is either too hot or too cold. A worst-case scenario would be having hospitality on the other side of a flimsy partition to the engineering office. You would be trying to concentrate during a meeting, but you are hungry – then the smell of food comes wafting through.

In Mexico every year, a bird would manage to break into the hospitality area and cause chaos. It was rowdy enough in Mexico when we were running Sergio 'Checo' Pérez. His supporters would be enjoying themselves in hospitality, making it necessary to wear a headset just to hear yourself think.

Although not a flyaway, Monaco is difficult to work at. The schedule at Monaco used to differ from other grands prix in that free practice was on Thursday, leaving Friday free for meetings and doing analysis. But, unfortunately, there was practice for support races on Friday, and the F1 team offices are right by the track. There was an outside chance you might be feeling a bit delicate after the night before, you were stuck in an office – which is hot because of its position above the garage – and racing cars were blasting past a few metres away on full throttle.

I found that, even though there was no track activity on Saturday night, there would be no peace when the music at nearby Rascasse kicked off. It was difficult to know what to do. F1 has a Parc Fermé rule which runs from the beginning of qualifying until the start of the race. During that time only routine work can be carried out on the F1 cars. With a few small exceptions, the specification cannot be changed in any way. That means team members would leave the track at a reasonable hour and return to the hotel. I tried going back with them and finishing the strategy work in my room – but that tended to be a bit of a faff because room service is all over the place, the food is not great anyway and, all the while, the guys at the factory are waiting to hear from you. It reached the point where I would stay at the track on a Saturday night. That was all very well, but then you would find that some of the crew would go into the bars and clubs rather than go back to the hotel. I would volunteer to bring them home once I had finished – which meant rounding up a load of rowdy mechanics at some late hour in the midst of Monaco in full swing.

As I said, Monaco is the exception when it comes to working at the European races. Usually, the strategists would work in the truck containing the race office. There, you would find each race engineer paired with their respective performance engineer, with the driver sitting in between so they could talk together. The strategist would usually sit with the chief race engineer, or the reliability and the systems people and the tyre engineers. When in a briefing, everyone would use a full intercom, which would allow the appropriate people back at the factory to join in.

If I was at home, I would work with music; the same on a flight. But in the office at the track, I felt it was important to pick up on conversations that happen around you. Someone might come in and say, 'The fuel pump isn't working,' or refer to something that has happened in the garage. The strategist might not be high enough up the food chain for people to pass this information directly to you – but it does affect your plan for the next session, particularly if you hear them say, 'It's going to be a bit tricky getting out [on time for the next session].' It is good to pick up on that early and then warn the guys at the factory that we need to be thinking of alternative run plans or, if this car does not go out, decide which tyres will be given back (I'll explain tyre allocation in the next chapter).

It is a matter of ticking off these necessary conversations. The engine guys sitting behind you might be having a chat about something totally unrelated to your car; it could be to do with another Mercedes engine. It might seem very minor but, in the overall scheme of things, because I had listened to the Mercedes radio discussions gathered during the previous race, I might be able to add something to that discussion. Keeping an ear to the ground keeps the strategist in tune with the little things that might eventually matter.

I felt that I was a strong strategist because I had experience of so many other roles. Having done my performance engineering at McLaren, I understood an element of the fuel requirements, or the brakes, or perhaps engine modes. I understood the fundamentals of what other people in the group were trying to achieve. I had the view that it would be better to have a little bit of information about many different things rather than just focusing on the basics of strategy. A strategist may know on which lap to make a pit stop (or 'box') but, if that does not apply to all the other departments and the way they are working, then it is no good. I always tried to take a wider view.

There can be an overlap in roles at the race track. For example, on race day, you are trying to put as little fuel in the car as possible – that is a performance engineer's role. But the strategists will get involved by saying how many laps in traffic we think we are going to have. Are we going to be one lap down at the end of the race? How much are we going to push the tyres? Is there going to be a lot of lift and coast (LICO)? LICO is used to save fuel, tyres or manage temperatures by lifting off the throttle early and coasting, rather than braking, into a corner. As a strategist, you end up involved in a lot of conversations that are not strictly strategy, but which could affect the result of the race.

The question of whether the car will be lapped by the leader is important because it means your driver will be doing one lap less in the overall race distance – thus presenting the opportunity to carry less fuel. That could mean carrying, say, 2 kilograms of fuel that is never going to be used. As a rule of thumb, if 10 kilograms of fuel is worth three-tenths of a second, then, on average, this could save 0.06 seconds a lap. Across a race of 60 laps, that is 3.6 seconds. Which is significant in F1 terms. Carrying less weight will also affect tyre wear and the launch off the start line.

If you do not know whether you are likely to be lapped, deciding on the fuel load becomes another example of risk and reward. It is one of the many things you are trying to manage in race strategy. If, for example, in the closing stages, the leader is not far behind and you know your fuel is marginal, there will be a discussion about instructing your driver to allow himself to be lapped.

The flow of information, conversation and decision-making continues throughout the race as a number of people with specific roles work together. In isolation, these seem small factors, but they all need to be considered. A strategist will be looking at tyre degradation while, at the same time, the tyre engineer will be keeping an eye on thermal effects with the aim of trying to establish how many laps this particular set of tyres will last. The performance engineer will be examining car systems to see if there is anything that needs to be done. The conversation can include several people that are not necessarily examining the same data you are looking at, which means it's paramount to communicate clearly what's going on to the others.

If, say, we are suddenly trying to push to get a pit window on someone, we may have only two or three laps to do it. When the driver pushes, the tyre engineer will say the tyres are getting too hot and won't last. Meanwhile, the performance engineer is reacting by adjusting the brake temperature. At this stage, it is about the strategist getting everyone on board and making sure they know this is only for a couple of laps and not the ten laps originally planned. That's difficult, particularly if it's wet or conditions are changing quickly. You need to get everyone on the same page quickly.

As mentioned, most teams have volunteers that help out over a race weekend. Generally, they are people from the factory whose core jobs won't involve racing, but they have a very strong interest in figuring out what goes on over the race weekend. They might sit in mission control and clip bits of video to make available quickly at the track. Or they might listen to rival teams' radio channels and transcribe the chat. By having the chats nicely colour coded, it means the strategist can quickly pick up on things, or get a feel for the wider picture.

If the conversation they hear is really important to us,

because it involves a driver running close to ours, then they get straight on the intercom and say the driver in question is about to box – which obviously could seriously affect what we do in response. The transcripts also provide a record which can be examined after the event.

This intercom channel was one of 14 I had on my main panel. I had channels specific to each car. Among the individual channels, there would be one for strategy and a channel for performance engineers to chat on. There would also be what you might call a 'high level' channel for talking to others on the pit wall or upper management. Each race engineer would have a channel and the strategist would have a means of communicating through the intercom with individual functions such as tyres, reliability and car crew. There is so much going on that I would not listen, for example, to the performance engineering channel. I would choose the channels that were very clear to my function. Anything important – such as discussing an upcoming pit stop – might be escalated to the higher management.

At all times, it was essential that the communication should be brief and relevant. That could be difficult because people have an unconscious habit of becoming commentators. The intercom needs to be very regimented, particularly if team members are fairly new and they feel there is a need for them to show they know what's going on when, in fact, they're stating the obvious and it's better to stay quiet. If, for example, the safety car is called, a warning will appear on everyone's screen. There is also an audible warning about the safety car programmed into the intercom. Despite all that, someone, somewhere, will still say, 'There's a safety car.'!

If the safety car is covering an incident involving a stationary car or debris on the track, it is sometimes useful to have feedback about things that are perhaps not on the strategist's data. A member of upper management watching the TV may comment on an umbrella going up in the grandstand – something the strategist has no means of knowing about. You do need people in positions we used to refer to as 'capacity seats'; people that are not flat out working to their maximum and who, like viewers at home, comment, 'Why are they not changing tyres? It's as clear as day it's starting to rain!' But, in fact, because you are so focused on the data, that something may not be as clear as day!

The potential appearance of the safety car is always tricky, particularly if you are coming up to a pit stop. With a bit of luck, someone will have already warned you that a car has stopped on track. Do you wait for another lap and hope to pick up the safety car? You will always get lots of different opinions across the intercom about that! It is possible the team's sporting director might have some information from race control, but maybe not, because the officials are probably overloaded at this point. This leaves the strategist with a difficult decision that must be made quickly, although much depends on where your drivers are on track relative to the pit-lane entry.

There is a risk in aborting a planned stop at the last minute. Either it's too late and the driver is on his way in and has crossed the white line defining the pit-lane entrance. Or the tyres are already in the pit lane, out of their blankets and cooling down. It's very likely that someone in mission control is listening to the radio communication between the driver in trouble and his team. You would then be told that the driver can't get the car restarted, or they're trying to find a default switch; or someone might say that the car is parked off line and there might not be the need for a safety car. While all this is going on, the strategist will be returning to the familiar topic of risk and reward. Do we absolutely need to box this lap, or was this lap chosen for some arbitrary reason? If it turns out there is no safety car, is one more lap going to kill whatever had been planned?

Another factor is that moments like this can affect your pit window. The situation can change so rapidly because your driver might lose a bit more time than your rival going through the yellow flag zone. This needs to be monitored, right up to the point where your driver is approaching pit entry. If the original plan was to box and come out ahead of someone then, because of the incident on track – and barring some miracle pit stop – you are no longer going to make the track position, it's probably not worth making the pit stop at that stage. You have to be reactive and constantly re-evaluating your options.

The ability of one team to listen to another's pit-to-driver radio has an interesting history. In 2016 it was decided that the teams were giving their drivers too much information. The FIA brought in a rule that said conversations with the driver could cover weather or track condition – or anything safety related – but very little else. The officials were putting a sharper interpretation on their rule that says the driver must drive the car unaided.

As you might imagine, this being F1, the teams came up with codes in which, say, a particular reference to the rain might mean something else that had been agreed beforehand. At the time, we did not have access to rivals' radios, which meant people spoke freely. But if your conversation got picked up and played out by the TV channels, then everyone got to hear it. The teams listened to these increasingly ridiculous codes and soon worked out they were all playing the same game. That's when the tell-tale complaints about codes began!

It did not take long for the FIA to decide, 'We can't be bothered with this any more. You can say whatever you want to your driver. But everything that goes from the race engineer to the driver is available, live, to the other teams.' This is brilliant because, on TV, you can follow onboard an individual car of your choice and have all the radio content available.

When it comes to radio communication, teams work in different ways. On my panel, I ended up with as many as 16 buttons that were linked to individual drivers from rival teams. I had them turned off most of the time because there was too much chatter. But occasionally, if you were racing someone closely, I would have the relevant channel on so that I could pick up as much information as possible.

The risk is reacting to something you think you heard – but didn't, because of interference. At one race in 2023 McLaren tried to dummy another team by saying, 'Box to overtake.' It was clear that that was not going to happen but, in the heat of the moment, you might think otherwise. You must either ignore data or question data quite thoroughly before you trust it. That's actually quite difficult in the time available.

F1 is constantly evolving. Strategy used to be based on a full lap time split into three sectors. You would note that one driver was quicker than yours in one particular sector and you would try to narrow down where you were losing out – and by how much. And that was more or less it. If that sector was high speed or low speed, you had a feel for where you could improve, relative to your competition.

With the increased use of GPS, it's possible to reduce the track to individual corners and discover more precisely where you are losing time. In the race, with a combination of the radio comms and the GPS, you can track your car and one you are competing with. By overlaying your rival's speed traces, you can see them 'clip'. This is when his speed at the end of the straight has dropped, indicating 'clipping', where the driver has come off the throttle, or perhaps been put on a different engine mode, to save energy and build up the battery. When you see that, you can warn your driver of an imminent attack on the next lap.

You get to know your competitors. In 2023, for example, it was noticeable how Alex Albon managed to have cars stuck behind him quite regularly because of the straight-line speed advantage of his Williams. I would have had someone in mission control delegated to analysing where others had overtaken Albon by checking the videos and noting rivals had got past at the exit of Turn 3, or wherever it might be. It is all about how much information you can gather – and how quickly. Then analysing it and making a decision everyone is waiting for. The outcome of your team's race depends on the strategist's decision being the correct one. No pressure, then . . .

Chapter 14

TYRE TALK

I t's fair to say that the subject of tyres will creep into every advance conversation about the race weekend. Sometimes tyres will dominate the discussion.

Thinking about a normal grand prix (as opposed to a weekend with a sprint race), we must decide, usually in the pre-event meeting at the factory, which tyres we intend to run in each of the three free practice sessions and qualifying. This is a good point at which to outline the weekend tyre allocation, as mandated by the FIA.

Given that there are always three different dry weather tyre compounds (soft, medium and hard) at every race, each driver is allowed 13 sets in total (eight soft, three medium and two hard). Because it is necessary to give back two sets at the end of FP1 and, again, at the end of FP2, teams usually run two types of tyre in each of these one-hour sessions. It's not obligatory – you could run three if you wished – but two is the norm. The same applies to FP3 on the second day.

The strategy team will have been looking at two main

things when building a tyre model: historic and recent data. For the first, we want to know how the track in question has performed in the previous two or three years. Is the tyre degradation very high or, perhaps, reasonably low? For some circuits, the combination of track temperature and roughness could lend itself to a soft tyre performing really well. But then it is necessary to note that the forthcoming race may be at a different time of year, which means the track has been much hotter in the past, or the wind could have been in a different direction or, perhaps, the track has been resurfaced. It is necessary to be really critical when looking at the source of your historical data.

Adding in the year-to-date figures helps build a matrix of specific elements relative to that track and how the tyres have performed in the season thus far. Pirelli might have developed a new soft compound for the current season. Has that compound been performing better or worse than last year? Do we understand the reasons why? If that soft is much faster, but with increased degradation, this will be factored into the tyre model and help build a picture for the weekend ahead.

Meanwhile, the tyre engineering group will have a more scientific way of doing that based on tyre energies. They will look at what they think the tyre, in theory, should be able to do at the circuit in question. By attempting to merge the two schools of thought based on statistics and theory, you hope to end up with a model that keeps everyone happy – well, reasonably happy!

The senior Queen's University Belfast Formula Student team of 2009. Taken on the old start/finish straight of Silverstone circuit.

Both of the Queen's University Belfast Formula Student cars at the 2009 competition in the garages at Silverstone.

In the garage for McLaren Racing during the German GP 2012, working as a reliability engineer.

A promotional photo for the UK government's 'Make it in Great Britain' campaign, taken next to cars on McLaren Technology Centre boulevard.

With Jenson Button in the McLaren garage as a performance engineer during the 2013 season.

Working as a system engineer on the McLaren GT project.

Left: With the team celebrating my first podium with Force India in Russia 2015.

Right: With the P3 trophy from the Russia GP 2015. Taken in hospitality after the race, as others in the background are avoiding pack-up.

On the Force India pit wall before Friday practice for the Brazil GP 2015.

On the Racing Point pit wall deep in discussion with Esteban Ocon during Saturday for the Monza GP 2018.

Image of SBG Race Watch strategy software used for both live analysis as well as simulations and planning. Top left shows live TV footage or could show car onboards. Top right shows track circuit with GPS positions of each driver and the gaps between cars. Middle right shows driver pit window centred on a driver. Edge of red section represents the edge of the pit window. Yellow bands indicate safety car pit window. Bottom right shows a driver's lap times against the planned lap times. This allows strategists to see if the driver or tyre are performing better or worse than modelling. Bottom left shows weather observations including ambient and track temperatures, wind speeds and directions.

On the Racing Point pit wall post-qualifying at the Hungary GP in Budapest 2020. Celebrating (through Covid) a second row lock out (P3 and P4) for the race on Sunday.

Post-win celebrations with Ryan at the podium in Bahrain 2020.

In the pit lane at the Silverstone GP 2021.

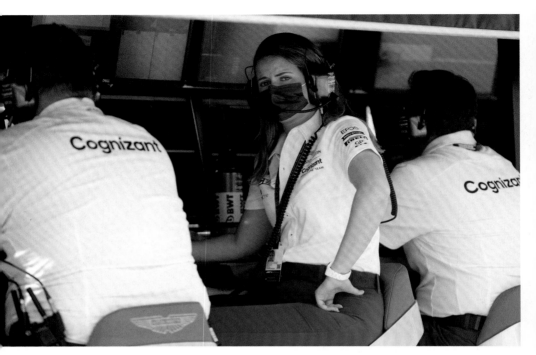

Working as Head of Race Strategy on the Aston Martin F1 pit wall during Saturday at the Silverstone GP 2021.

My team headshot photo for 2021. Taken during the 2021 Brazil GP weekend.

My headset back in position for a final time after the Budapest GP 2022.

Morning of the final race weekend with Aston Martin F1 for the Budapest GP 2022. I am pictured in front of the garage alongside Sebastian Vettel (left) and Lance Stroll (right).

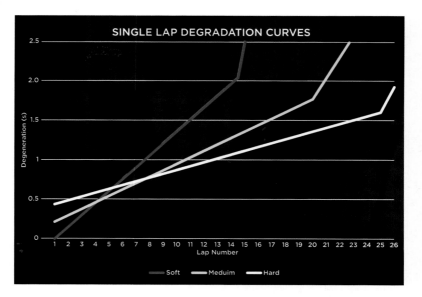

The single lap degradation curve chart shows the lap time for each compound vary through the stint. It shows that on lap one the soft tyre is quickest. It shows that on lap four a medium becomes quicker than a soft as the lines cross over. It shows that the life of each tyre is exceeded when the lines sharply increase.

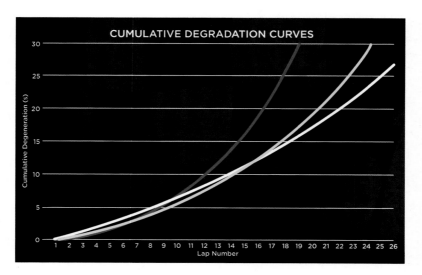

This is used with the previous chart to see how each tyre performs over the entire stint. Using the same base as the previous chart, even though the medium tyre was quicker than the soft on lap four, it is not until lap seven that the cumulative chart crosses over. That means that for a seven lap stint a soft is quicker than a medium tyre as the time benefit at the start is greater than the loss on laps five and six.

Accurate modelling is made difficult because tyres are very nonlinear in their behaviour. This, in some ways, goes against the grain for engineers, who tend to focus on tiny differences of 0.01 of a second or 0.02 seconds when, in fact, it probably has little effect when creating a tyre model. There is a compromise to be struck because, basically, this is no more than an estimation of tyre degradation.

Estimating car pace is tricky. On the strategy side, rather than predicting absolute lap time, we are trying to establish where we sit relative to others. Our lap time could be a little bit in or out. That's okay, because this is all about where we sit; how much faster than us do we think Red Bull will be? How much slower will Williams be?

It's all about establishing where we think we are going to qualify. Will we reach Q3? Do we therefore need tyres for Q3? You're trying to answer some big picture questions. At this stage, it's not about determining absolute strategy, but thinking about the possible need to save a soft tyre for Q3. Will our pace be so good that we can run only a single set of softs in Q1? All of this is part of building an overall picture of how fast we will be and which tyres we will need – and when.

If the race is going to be difficult in terms of tyre management, do we think tyres will need to be saved from very early in the weekend? At this stage, the main question is determining which tyres we might need after Friday. The worst-case scenario would be to use all your additional hard or mediums on Friday, discover degradation is really bad, and then realise you have no flexibility of choice for the harder tyres on Saturday. Answering such big questions is more important at this point in pre-event planning than getting caught up in a conversation about, say, what to do if there is a safety car early in the race.

From 2016 until 2020, you had to choose your mix of tyres and tell Pirelli eight weeks in advance. That was another day's work that the strategy department didn't really need, as you tried predicting variables such as car pace and the weather, and how they would be in eight weeks' time. That may have been bad enough, but it was even more difficult for the first races of the season when 13 weeks' notice was required to allow for your choice to be shipped to Australia and the Far East. In the days before Christmas, you would be making an educated guess at the pace of a car and tyres, all of which had yet to run and were completely unknown. The discussion and choice had to be based on nothing more than estimates.

My personal choice is the allocation of sets outlined previously, because it is the same for everyone and you adapt to suit. It could be argued that the same allocation for every race has its downside, particularly for Monaco, where no one wants the additional hard or medium tyre, full focus being on qualifying.

Which tyres you use – and when – depends on your expectations during qualifying. Teams in the lower order of the championship would not have a realistic expectation of reaching Q3. It would make sense, therefore, to run two sets (of the five) soft tyres in Q1; two more in Q2 and have one set left if, hopefully, they reach Q3. The front runners, of course, would assume no difficulty in making it through and may not need any soft tyres for Q1.

Bearing in mind that each driver must use two different compounds in the race (assuming it's dry), life can become tricky if two pit stops (or a 'two-stopper') are expected, most likely using the medium and hard. On this assumption, if you have an additional medium or hard for the race, you will only have four softs to get through qualifying: two for Q1, one for Q2, and one for Q3. So you end up in this very difficult situation where the soft that you run in Q1 is very important: do you then need to run the second one or can you save it for Q2? And then comes a compromise between wanting to get to Q2, but also wanting to have two softs when you get there.

We're getting ahead of ourselves. Going back to free practice, FP2 (on the afternoon of the first day) is generally the most representative for qualifying because it is at roughly the same time of day. Naturally, you will want to run a qualifying tyre (soft) in FP2 but, because you are generally limited to a qualifying tyre and one of your race tyres (either medium or hard), it may be necessary to decide whether to split this across the two drivers.

There is also the question of the weather: is there a risk

that FP3 is going to be wet? In which case, should you try to move the tyres to different sessions on the previous day? Monaco is an outlier. Because qualifying (and grid position) is so important for a race in which overtaking is almost impossible, you will try to run a soft tyre in each session so that, come what may, you have at least got the experience of a qualifying (soft) tyre at some stage.

As you can see, much of the decision-making is based on the characteristics of the track, be it difficulties in over-taking, the prevailing weather, the likely track improvement and the effect on the track condition of practice (and races) for support events such as Formula 3 and Formula 2 or GT cars. Taking all these considerations into account, a plan is formulated, the aim being to have everyone in the team buy into it at the start of the weekend.

Then it becomes a matter of sorting out the detail, such as needing to have experience of the medium and the hard tyre during FP2. In a perfect world, the drivers will have different preferences. But that rarely happens. It will come down to the toss of a coin (or whatever system of choice a team may use) with the inevitable result that one driver will not be particularly happy.

Information on both tyres is needed because the strategist is continually trying to predict a tyre model. To do this, it is essential to establish as best as possible the steps in lap time between the soft, the medium and the hard tyres. You also need to know the lap time at low fuel, and at high fuel. (I will go into detail on how this is established during my explanation of what happens in each practice session.)

The second critical element is establishing the degradation on each tyre compound; in other words, how much the various tyres will become slower on each lap. Throw into that mix the difference between the car running with low fuel when drivers are pushing really hard, and high fuel when they are pushing more consistently. The final element of the tyre model is working out how many laps each tyre will be able to run before the performance seriously diminishes (or 'falls off a cliff').

Apart from building a picture, the strategist is also trying to align people and channel their thoughts about what to expect.

Teams will spend a lot of time beforehand in the simulator – how much time depends on the event. New tracks are the most difficult because you don't have any historic reference. You can form a tyre model on one compound and then make the standard performance steps between this compound and the other two. You're also trying to find the data in other ways.

Let's say a hard tyre for the events thus far in the season has generally had half the degradation of a medium tyre. Using that as a basis, you start to formulate the trends between tyres. But you need a long run, let's say in the simulator, to roughly work out the tyre energy being used over a lap. Then you equate that to other tracks when trying to establish a baseline model. There will be a lot of interpolation between data sets. It is a non-stop process, and the race hasn't even started yet . . .

Chapter 15

OPTIMUM PLANNING

The tyre model is the foundation for all strategy. Key elements such as the pace comparison between the three different tyre compounds, degradation and the life of each tyre, are built into this model.

The other factor is the base lap time. This would be arrived at by assuming zero fuel and zero degradation. Plus considering the number of laps in the race and the time taken to do a pit stop. In that pit-stop calculation, we would include the time taken to come into the pit lane, the time spent in the box changing tyres, followed by the time taken to leave.

We would allow between 2.5 and 3 seconds for a pit stop, which may sound a lot when you hear that pit stops can be as fast as 1.8 seconds. We would choose a middle ground. There are many things to consider, such as the stop itself possibly being fractionally slower than hoped, plus the fact that the 1.8-second or 2-second stop does not include the driver reacting to the lights and launching from the pit box. Neither does it take into account the driver messing up the entry or being slow for whatever reason. This is why we would allow a small margin over and above what you might expect.

Taking the tyre model and the race distance, we would work out the fastest race for an individual car. In other words, if you were the only car on the track, what would be the fastest time taken to complete the race distance, and how would you do it? Would it be a one-stop or a two-stop race? Is it using the medium tyre and then the hard tyre? What is the best combination? This would be a mathematical model establishing the quickest way to run the race while respecting the regulations governing the need to run at least two different tyre compounds, which must include the medium or the hard.

Continuing with the single-car theme, we would look at all the alternatives. How slow is a two-stop compared to a one-stop? Looking at a one-stop race using, say, the soft and the medium, how much slower would it be if you ran the medium and the hard? We end up with a matrix of combinations showing the difference in race time between each variation.

We would also know, for example, the difference in race time between stopping on, say, lap 18 as opposed to lap 20 in an optimum one-stop race. In this case, you would save some degradation on the first set of tyres by coming in two laps earlier but, on the other hand, you would add two slower laps at the end of the second stint. One way or another, you establish the fastest race for a single car, and this is the starting point when establishing your race strategy. Then we set that aside.

The next step would be to examine a 20-car race, with the first questions being: where are we likely to qualify? How much faster or slower are we compared to the cars around us? How difficult is overtaking going to be on this particular track? How far is the field likely to fan out once the race is under way? What will the spread be after, say, ten laps?

Establishing a likely pit window comes next. Where would we rejoin the field after a stop when considering the likely positions of the backmarkers? When are the leaders going to stop and come back onto the track? It is a case of trying to work out what the others are likely to do and how it would affect your optimum strategy. How do we achieve, not necessarily the quickest race, but the best finishing position based on interactions with others?

The biggest factor in all of this is overtaking. At some-where like Monaco or Singapore, you will do a sub-optimal race in order simply to retain or gain track position. This means either making one stop less or stopping at a different time than the quickest race strategy suggests.

Track position is all about being in front of another car, either through having started ahead of them, or having undercut them and moved in front during your pit stop. Track position is key. We end up with races that largely fall into two categories: one in which track position is important – those are the races where it's difficult to overtake; and one where optimum strategy is important.

Bahrain is a good example. This is an optimum strategy race because overtaking is comparatively easy and degradation is high. Here, you want to make your pit stop as close to that perfect race plan as you can because you should be able to overtake other cars thanks to having fresh tyres. On other tracks, that's not the case.

In an optimum race, you will end up with a plan that is very, very close to the single car race. In the races where track position is very important, you end up with a plan that's often very different to the optimal race. It could mean starting on a tyre that is different to the normal choice. Singapore is a good example. Even though starting with a soft tyre is the best choice for the fastest race, you might start on a medium tyre in order to allow yourself to go longer into the race in the hope of picking up a safety car and the chance of a pit stop with less time loss.

In some races, you buy yourself flexibility with the strategy you choose. If there's a good chance of a safety car, or a higher risk of traffic behind you and you don't want to be forced into an early pit stop that would mean rejoining into traffic, you need the flexibility to extend your first stint. You see that from the elite teams as much as you see it from the teams lower down the order.

Equally, a soft tyre at the start might not be the quickest

strategy but, if you want to gain positions off the line, knowing the start is your best overtaking opportunity of the entire race, then the soft tyre could be the best choice. The fastest tyre at that point might make a big difference to your position later on. On the other hand, being on pole and not wanting to lose positions off the line could mean choosing the fastest tyre in the beginning in the knowledge that you can drive a bit slower and manage it later on. Many of these external factors are being weighed up when putting the race plan in place on race morning.

Different tracks may have individual requirements but, as a rule, track position, or being ahead of a car that's slower, is the obvious preference. Running in free air brings many advantages. You can go at the pace that suits you best. That might be as fast as a car can go, but that's not always what you need. Running the car a little slower than maximum pace could help manage tyres or fuel. That becomes easier if you are out on your own because you can run the race as you wish and stop when you're ready.

Being in traffic is very different: there are many consequences, some of which are good, some not so good. The most fundamental penalty of running in traffic is that it changes air flow to the car which, in turn, affects performance, particularly when cornering. Disturbed air flow reduces the level of downforce. That may be an advantage on the straight by increasing maximum speed. But it hurts when cornering.

This is less critical at tracks such as Monza, where the straights are far more significant than the corners and explains why drivers tend to focus on getting a tow from the car in front in order to maximise straight-line speed. Elsewhere, running in traffic can be detrimental. The associated loss of downforce reduces loading on the tyres. This means, when cornering, the tyres will slide rather than grip the surface and not only lose time but also increase the degradation.

The reduction in air flow caused by running in traffic also affects temperatures across the car, from brakes (and, by association, the surrounding tyres) to the engine, radiators and oil coolers. This can be particularly detrimental if the car, starting from pole position, has been set up to run in the ideal situation of clear air and the driver finds himself, for whatever reason, stuck in traffic and not enough allowance has been made for cooling. That is why it is necessary to spend a lot of time from a strategy point of view predicting how much of the race will be run in traffic, because this influences the temperatures and performance.

This can be particularly critical in Mexico, a high-altitude circuit where cooling is right on the limit and the track layout makes overtaking difficult. I have been in situations where we missed the pit stop and stopped a lap too late. As a result, our driver found himself in traffic. The rest of the race was spent pushing hard for a few laps in an effort to overtake, only to have to back off to cool everything down. That became a cycle that seemed to go on for ever. We made no progress at all. But neither did anyone else in that bunch of cars.

A midfield team will spend a lot of time in advance of the race weighing up all the factors that influence strategy decisions. A lot of the choices will be focused on trying to avoid unnecessary interaction with others, and could lead to not running what might be the optimum strategy.

For example, 15 laps into the race and close to where you want to stop, it becomes obvious there is traffic that you need to avoid when your driver rejoins. There are a few ways of doing it. If you are four- or five-tenths of a second faster than the car you wish to avoid, you can calculate how many more laps you need to complete before stopping with enough margin to get out ahead and clear that car. Or you can wait until that particular car stops and removes the problem by no longer being in your pit window.

At this stage, you need to be listening to their radio conversations in the hope of discovering when they are thinking of stopping. Ideally, having enough pace to clear that car is the best solution because you have more control without relying on what the other team may or may not do. Generally speaking, teams at the back tend to be unpredictable and sometimes fail to use the smartest strategy. Knowing which tyre they're on can make a big difference. If they've started on a soft, they're likely to stop earlier whereas, if the initial choice was a hard tyre, they are almost certain to go longer.

If you can't clear the window of the car in question then, because of having a faster car and fresh tyres following your stop, it becomes a case of having the theoretical pace to overtake them in any case – provided the circuit allows it, of course. The strategist will arrive at a lap time they think is needed to allow this to happen, while also taking into consideration backmarkers (such as Williams in recent years) being very strong in a straight line and possibly more difficult to overtake than someone else who's running exactly the same lap time but is slower on the straight.

Part of the strategist's thought process will involve talking to your driver or your tyre engineer and saying five more laps are needed to clear the car in question. Can we do it? The driver might say five more laps are possible, but he needs to point out that the pace might drop off (due to continuing tyre degradation), so you will lose the pit window in any case.

In addition, you will be thinking about the possibility of a safety car window when fewer positions will be potentially lost through a pit stop because of the reduced pace of the cars on track – but, critically, not in the pit lane. This means that the time lost relative to someone on the track changes significantly, usually to about half the amount of time.

Under normal circumstances, you might lose 25 seconds in a full pit stop – by that we mean returning to the track 25 seconds further back than you were before coming in. This will have been indicated on our pit-wall screens, the expected position for the return being shown by a 'ghost' marker which follows our car.

Under a safety car, however, the time lost might be just 12 seconds, with the accompanying loss of fewer positions on the track. The strategist is constantly looking at the position of other cars in the event of a safety car; basically, who is in our safety car window? If there's a safety car, who (that we may otherwise have beaten) is going to gain over us? If we don't need to stop on a particular lap, what are the repercussions of stopping and then another safety car appears later at a point that might suit us better?

If you're racing someone very closely, you can't delay your stop indefinitely in the hope of a safety car situation because you don't have the pace or the tyres to do it. You are worried about them stopping one lap earlier than you or having the speed to overtake. All of which forces your hand into making a pit stop. If you do have the margin, you're trying to stick to the optimum race because you know that's the quickest.

When attempting to undercut the car in front, you will be looking at your pit window and checking everything in terms of who you would come out behind and how long it will take to clear that car. That becomes more critical when there is a car directly behind you, or you're directly behind someone else because you're trying to undercut. The car ahead will clear the pit window first, meaning they will be taking less risk with whatever pit stop they do. Sometimes the car behind must take a bit of a risk to undercut because the car ahead should have a clearer window than you do because they've got that extra three or four seconds.

Choosing when to stop is quite intricate and it is essential to keep the driver abreast of what you are thinking. Each team is doing this, of course. The strategist might be listening to whichever team is in close company, but there is no opportunity to monitor the rest, which is where mission control comes into play, as they listen for certain things that happen in advance of a pit stop. You might ask the driver about the balance of the car: always a tell-tale sign that you're considering a pit stop. Or you might have told your driver about opening a safety car window – another giveaway that a pit stop could be close.

As mentioned in the previous chapter, teams can use codes, some of which are more successful than others. For many years, Haas used 'Mode T4' so often in advance of a pit stop that it became obvious. Or you could go to the trouble of trying to hide your intentions and then the driver replies, 'Yeah, box this lap,' rendering the code pointless. Either way, the pit-stop decision is key and often leads to following a strategy that is not the fastest on paper, but is the best way to keep vital track position.

Chapter 16

FREE PRACTICE 1 (FP1)

After all the discussions and pre-planning, we are finally ready to see some action on the opening day of practice. But first, there are other important decisions to be made. Breakfast!

For the European races in particular, breakfast in the hotels usually left a lot to be desired. The preference was to go straight to the track and set yourself up for the day with a decent breakfast, courtesy of team catering. Even though this would be about three hours before the start of practice, there was much to do and time would be tight. Which led to a dilemma – at least, as far as I was concerned. Do I go straight to hospitality? Or do I dump my bag in the truck first – and risk being caught up in a discussion with the engineers that could go on long enough to rule out finding time for breakfast?

On the way into the track, I would be checking the skies, hoping for obvious signs of a clear day ahead.

One of the most difficult challenges for strategy at this stage is uncertain weather and the tricky question about needing to prepare intermediate or full wet tyres should rain be likely.

One good thing in favour of rain tyres is that they only need to be in blankets for an hour before the start of practice. Additionally, in the regulations for 2023, tyre blankets were totally removed from the full wet, but not the intermediate, simplifying the problem even further. Heating slicks takes much longer, and the problem in the event of a change of plan is not so much turning wet tyres on, but pulling the plug on dries that started the heating process when the mechanics and tyre technicians arrived at the track. Much depends on the time factor, but it is not ideal to heat up a tyre, only to let it cool down. The extra heat cycle means the tyre will no longer have the ultimate performance thanks to not being as soft as it once was.

This potential difficulty is not so acute at somewhere like Bahrain, where overheating is the main problem. Curing the tyre in this way might actually be better for overall lap time. Much therefore depends on where you are in the world but, either way, the strategist must decide on arrival at the track – even before breakfast!

Checking the weather radar for rain is obviously essential, but this is also the moment to see if ambient temperature, wind conditions and track temperature are as expected. A tricky factor to deal with at this stage is cloud cover, because this is difficult to predict yet can have a massive effect on track temperature.

The last briefing for FP1 will involve using intercoms in the truck, about an hour before the session starts. There will be a summary of the main things that we need to learn, and anything that may have changed since the previous briefing. The final 15 minutes will be spent getting into position on the pit wall, sorting out any final niggles, checking we're good to go – and perhaps realising you need to nip back and get a coat! Headsets would be plugged in and radio checks carried out between those at the track and colleagues in mission control.

The initial period is spent becoming acclimatised to that particular environment. The pit box will be in a slightly different position to the circuit before. As for the track itself, this might be anti-clockwise instead of the predominant clockwise choice, meaning the cars will be arriving and leaving the pits from the opposite direction. One of the first tasks is to check if your cars can leave the garage in one clean sweep, rather than having to be rolled backwards briefly to avoid the pit wall. This may seem relatively unimportant but it can have a massive effect on your calculation of the release time needed to reach the end of the pit lane.

On the same subject, the strategist will use a GPS map to establish the speed of cars approaching on the main straight. By checking where a rival is relative to, say, the final corner and then seeing where he ends up as your car leaves the pit lane, it is possible to establish the pit exit time needed to have your driver exit the garage and join a reasonably clear track. We add these garage release lines to our GPS maps and share across the team, so everyone is aware of the optimum time to send the car out on track.

There could be anomalies at circuits such as Baku and Monaco where the surrounding high buildings cause little glitches on the GPS to the extent where, until it becomes familiar, you have the impression that a car has stopped on track when, in fact, the GPS has frozen briefly. And each time the car returns, it is the first opportunity for the driver to practice stopping on the precise marks needed for the pit stop.

FP1 is largely a session of change because the track condition shifts considerably during the 60 minutes. It becomes quicker, perhaps because it is dirty initially through lack of use, but certainly due to the drivers getting their eye in. FP1 lap times can be reduced by two or three seconds; maybe more. This makes it a very hard session to learn anything significant from a strategy side.

FP1 is often used for testing aero or mechanical parts, whether small or significant. A lack of between-race testing means that FP1 is an opportunity to run a new component, with onboard sensors providing the answers rather than lap times. It is also an opportunity to try different setups across the two cars. It could be that during the previous race weekend, one car experienced higher tyre degradation or, perhaps, a poorer balance than the other car. The overall aim, however, is to have the cars set up ready for FP2, but knowing that the underlying problem with FP1 is the results proving very hard to read because the track improvement is so high.

The benefit of running high or low fuel is limited for the same reason, plus the track temperature will be different because FP1 takes place three or four hours earlier than qualifying and the actual race. Attempting a qualifying lap is also tricky because it is impossible to answer basic questions such as which is the best sequence of laps: is it a fast lap (or push lap) followed by a cooling lap before pushing again? Is the second push going to be fast enough, or is it going to be too slow? Do we need two cooling laps (a 'double-cool') in between? It is a struggle to find an answer in FP1 because the track and temperatures are changing and improving so much that the second push lap is going to look good even though the reality may be quite different.

An additional unique element to FP1 is the appearance of novice drivers. This session is the one opportunity for teams to have their junior drivers experience the urgency and routine of a live practice session. This will have been mentioned in our briefing to make everyone aware of new names appearing on the information screens. Those not paying attention will then be caught out by the three-letter acronyms representing the novices. Not only will they not know who these drivers are, but they will also have no idea which team they are driving for! But then, the strategist must ensure the software input is done correctly, otherwise that driver (or drivers) might not be represented by a colour, which, in turn, will create minor panic among some crew members at the beginning of the session. And, as mentioned before, this could be the moment when I would fail to handle the newcomer's name correctly and have that mispronunciation religiously passed down the line!

Naturally, the novices will be comparatively slow, at least during the early stages of FP1. The strategist will treat them no differently to the regular F1 backmarkers. When doing a long run, for example, every effort will be made to find space on track that avoids having your driver stuck behind, say, a Haas. Which reminds me of the theory that, rather than aim for the best bit of track, it makes more sense to find the second clearest section because everyone else is going for the obvious bit, which is why it becomes crowded.

While looking for a decent space, another objective is to place your driver in phase with someone on a similar strategy to yours – be that fast laps or high fuel running. Also, if the engineers want an aero test, they will need the car to be in a gap of at least five or six seconds behind the car in front otherwise the test will be negated by turbulent air. That takes a lot of management through the lap, and it's necessary to let the engineers and mission control know if the gap was good or otherwise.

Because various programmes are being run during FP1,

it is important to have everyone on the same page during what often feels like a very messy session. The planned programme can change quite quickly, particularly if the engineers did not get the answer to a test and the run needs to be repeated. This will immediately throw your run plan out of kilter and, while all of this is going on, the strategist will be trying to avoid the worst-case scenario, which is having your cars – the two cars you can control – tripping over each other on the same piece of track.

In many ways, FP1 is a test session. It is all about the driver – and everyone else in the team – getting up to speed. Information can be gathered about the effect of traffic and the cooling setup in the car, even allowing for the temperatures being different because these readings can be offset against what is expected in FP2. If you reckon the track temperature in the next session is going to be ten degrees hotter, then a number for that is added to the calculation.

The threat of a wet FP2 can throw a curveball into the plan for FP1, because this will be the only opportunity for dry running on Friday. Similarly, a red flag in the middle of the long run in FP2 could mean having to resort to data gathered in FP1. As a rule, I tended to believe my pre-event numbers rather than anything resulting from the inconclusive runs in FP1. Apart from allowing drivers to practise starts from the grid, the end of FP1 is also a good time for a pit-stop rehearsal as the crew gets the chance to practise with a car coming into the pits at the full permitted speed.

Once the session has finished, the priority is to debrief and discuss the conclusions reached. Various teams do this in different ways. When I was with McLaren, the debrief would be immediate. The drivers would climb from their cars and plug into the island in the centre of the garage, ready to talk to the engineers. At Aston Martin, we would gather in the truck for our meeting, probably about 15 minutes after the end of the session. There are pros and cons for both methods.

Information from the McLaren-style debrief would give the factory an immediate start at sorting out whatever important issues may have arisen, or the one thing that everyone agrees is going to find lap time. The downside was that there had not been enough time to pause, look at the data and verify the accuracy of the alleged problem. With the second option, the interval allowed a quick review by the trackside team and, at the same time, mission control would have already examined the data and perhaps have some suggestions ready for when the debrief started ten minutes later.

From the strategy side, I liked to arrive early for the Aston Martin debrief. I felt I needed to be ready because I'd be one of the first people to summarise the session, with details of a track improvement, competitor information, or perhaps some crucial traffic issue. There is a need for context. If one of the drivers said this second run felt a lot better, it was very important to know that, regardless of changes made to the car before that second run, it was always going to feel better because of the improvement to track conditions.

Whatever needs to be said, it is important to get through the debrief efficiently and quickly. There is a set amount of time (between two and three hours) separating FP1 and FP2. The longer the debrief takes, the less time available to do anything else before FP2. In any case, when the pressure is on, it is human nature to switch off when your section of the debrief is done. Throughout, people are beavering away on the next session. You may be trying to listen to what's going on but your thoughts are on timing or alternative software setup changes or whatever you believe could be done before the next session and, in truth, you want to get on with it – and maybe grab a bite to eat along the way.

That said, the debrief got longer and longer over the years, which led to various attempts to reset back to a shorter duration. Not everything needed to be said at that point, thanks to the subsequent issuing of post-session reports. I would do a strategy summary with various lap times, a description of the laps (low fuel, high fuel), our thoughts on track improvement, plus the sectors in which we were fast or comparatively slow. In simple terms, the debrief would be about getting across facts and thoughts that people might not find in the reports. Time goes very quickly indeed when it comes to having the cars ready for FP2 and knowing that something significant is about to happen.

Chapter 17

FREE PRACTICE 2 (FP2)

Free Practice 2 is a totally different session; there is a set order to it. On paper, FP2 is similar to FP1 in that the same amount of time is available and two sets of tyres must be handed back. But whereas the first session is about settling in, learning about setup and establishing what's needed, FP2 is much more regimented. With the time of day matching what is to come in qualifying and the race, you run a harder tyre (be that a medium or hard) for between three and five timed laps before moving to the soft tyre – in effect, your qualifying tyre – for up to three laps, which is generally the maximum for a soft at its best.

With the low-fuel runs complete, teams have different ways of moving on to long runs. Low fuel means somewhere between 40 and 50 kilograms of fuel, perhaps dropping to 30. It is not a qualifying level, but neither is it high fuel. For the high-fuel long runs, the load ranges between 80 and 100 kilograms.

There are many different theories about which level to choose – and when. Running very low means time lost when filling up the car (which must be done inside the garage rather than, as it used to be many years ago, in the pit lane) for the long run. Choosing the middle ground also presents an opportunity to check the car's balance at this level, which is important for the race. Monaco is the exception because running a low level is critical for qualifying which, as has been mentioned, carries more importance than the race, with its overtaking difficulties.

At the other end of the scale, some teams choose not to waste time going all the way to the maximum of 100 kilograms because that level represents only a small proportion of time in the race. Whatever level is chosen, a lot of time is spent watching other teams and working out an average fuel level to correlate their estimated pace to your own.

Red Bull and Mercedes always ran heavier fuel than we did at Aston Martin, which meant they were likely to be even faster on race day. As a strategist, you would try to average that out – taking into account probable engine modes and traffic on a fast lap – while getting a feel for where they might run in qualifying and the race. When looking for fast lap outliers, the software available now is very good at highlighting a rival experiencing traffic thanks to having the GPS gaps between cars. In the case of a driver making a mistake, that should show as an outlier slow lap, pure and simple.

Meanwhile, the strategist will be trying to figure out a number of other things relative to the race: tyre degradation is an obvious one; projected tyre life will be another. A limiting factor is the time available, since you will probably do between eight and ten laps in a long run during FP2, whereas it is likely to be twice that number for each stint in the race.

To get a good read across the tyres, teams might split the work by having one driver focus on the medium while the other concentrates on the hard. But that can be awkward because, on some tracks, the hard tyre might not be the preferred choice for the race and the driver running it will feel at a disadvantage, particularly in an important session such as FP2. Splitting the tyre running in this way can also be a compromise when it comes to gathering the necessary information because it is difficult to compare drivers; for example, you can't do an overlay across them because of the different tyres.

One of the fundamental problems associated with strategy is if two people attempt to work out degradation using the same data, they will probably arrive at slightly different answers. It is a case of arriving at a number you're happy to use as a baseline for later calculations.

As ever, the strategist needs to keep an eye on several things, particularly during the long runs. It could be that another car joins the track just as yours is about to complete, say, lap five of a ten-lap run. If your driver then says he is backing off to create some space, that slow lap is useless and ruins the readings for the entire long run. Similarly, if the driver suddenly chooses to use DRS to check out how effective the overtaking may be, that also corrupts the information coming from the long run. Or, if he pushes really hard on lap one and causes the tyres to overheat during the next eight laps, that really hurts, because your one chance of a useful long run is gone thanks to time running out.

Despite the danger of running out of time, it is preferable to have the long run at the end of FP2. The problem then comes if you think there might be rain at the end of the session. Or, in the case of Monaco, the high chance of a red flag as drivers hit the wall when pushing too hard as the clock ticks on. The temptation is to bring the long run as far forward as possible to gather the essential data – only to find that the rain stays away and the session runs trouble-free to the finish.

That becomes tricky if further running is restricted because the engine is becoming tight on mileage that weekend. And, of course, that will be when a driver aborts a fast lap and says, 'Oh, I'll just go again.' Okay – but we've just lost 3.6 miles, or whatever it might be. It is worth pointing out that this occurs throughout the season with its limit of four power units per driver. Mercedes, for example, would balance our supply across the whole season which would mean, even in the early races, there could be a mileage limitation in the interests of reliability.

All of this can be compromised further by a red flag bringing the FP2 session to a halt. On paper, this is not such a massive issue, the theory being that it's the same for everyone. Nonetheless, a red flag can be disruptive if, say, you are in the middle of a long run to determine degradation. Much depends on when the red flag happens to appear during the session. If it's in the second half and you chose to go out early, that will be to your advantage; compared to other teams, you've got a few more laps on the board. You may feel that you're ahead of the game because of the information gathered, although, if the session runs non-stop, then the late runs may have been more useful because the track condition has ramped up slightly – albeit a bit less than earlier in the day. It is a gamble either way.

Regardless of when a red flag occurs, the first question will always be: 'How long do we think the red flag is going to last and how many laps do we think we can get in when FP2 resumes?' As I mentioned in an earlier chapter, this is where mission control and anyone trackside with eyes on the TV monitor can help with news of what caused the stoppage and how long it might be before the problem is dealt with. It could be, for instance, that a car has come to a halt and simply needs pushing off the track. But if there's been a crash, the extent of the damage to the barriers will determine how much work will be involved in repairs and clearing the debris. If the delay is likely to be lengthy, the strategist must be aware of suggestions to make changes to the car. Resetting the fuel level, for example, would have a major effect on the resulting degradation calculation.

FP2 can be used to practise a pit stop or get a feel for the start performance of the tyre most likely to be chosen for the first stint. The drawback there is that the launch, which must be done on the starting grid at the end of the session, will be on worn tyres and therefore not representative of the real thing.

All told, the strategist is presented with this matrix of trying to figure out what is needed from each tyre. Ideally, you finish FP2 with good information on how to get the best out of your qualifying tyre (preparation on the out lap, for example), the ideal gap to the car ahead, and an indication of how the long-run tyres might perform.

FP2 is busy but, compared to FP1, generally feels a steadier session because everyone is more aligned. When drawing comparisons with other teams, you know that everyone will have run the soft tyre, or completed a long run, at roughly the same time. Cross-car comparisons are more consistent; traffic is better because everyone tends to be on the same run plan, even allowing for one car perhaps leaving the pits later than the rest because of some issue. The strategist feels more in control thanks to knowing when to release a car, predicting the effect of traffic and being aware of when it's a good time to speak to the driver. Everything feels calmer because you are not worried about trying to plan a pit stop or thinking about possible safety cars – as you would during the race.

FP2 is definitely the most useful session, when everything starts to come together after days seemingly spent dealing in speculation and unknowns. That hour allows most of the strategy and race engineering analysis to be done. Apart from gathering solid data on your own car, the strategist has also been able to establish field average data, checking out degradation for everyone else. That is important because ascertaining degradation on, say, the medium tyre for just one car is not good enough; you need to take the whole field into account to arrive at average degradation, track improvement and the performance step between the various tyre compounds. But, then again, I found taking readings across the board to be the hardest aspect of strategy compared to other brands of engineering – averaging does not feel good to an engineer!

Similar to the end of FP1, following FP2 there will be a debrief, but longer because there is not the same sense of urgency. The one time-restraint on a Friday is a curfew which comes into force four hours after the end of practice (having been reduced from five hours when I was racing). The time limit was introduced several years ago to prevent teams – as was the popular habit – from working long into the night. It was part of the motor racing ethic of getting the job done, no matter what. That was in the days of 15 or 16 races in a season. Now, with a possible maximum of 24, the curfew removes the temptation to work very long hours.

Details of necessary setup changes need to be relayed to the mechanics in time to have the work done before the curfew and ensure everything is ready for FP3. Even so, you'll find final changes being made on Saturday morning thanks to either the work not having been completed before the curfew, or as the result of analysis done overnight by mission control or those running analysis through the simulator. At the time of writing, there is no limit on the working hours at mission control to match the increasingly tight schedule for the guys at the track. During my final season with Aston Martin, Friday nights had reached the stage of fitting in meetings around doing your analysis, formulating a plan going forward and grabbing dinner at the track.

The central focus of the strategist's work at this point is to rerun the race simulations, this time with the latest information which, hopefully, will have eliminated a few of the previously unknown factors and increased your confidence. By the end of Friday night, you should feel more comfortable with the tyre model, be that knowledge of degradation or the pace gaps between the different compounds.

Overall, there should be a better understanding of how your car compares to others going into qualifying. This is important from the strategy side when it comes to predicting just how far your car is going to get through the various stages of Q1, Q2 and – hopefully – Q3. It will affect how many soft tyres you think are worth keeping for qualifying, and whether they should be used in FP3 on Saturday morning.

This is also when the plan for qualifying begins to take shape. How many runs do you reckon it's going to take to get through Q1? What is the best time to do those runs? Will a used tyre be any good, or is it simply not worth thinking about? How effective would a second push lap be?

Similarly, on Friday night the strategist is also trying to establish the number of tyres needed for the race – which obviously depends on the number of stops. If the race is two-stop, then an extra set will be needed and that must be worked into the calculations for FP3 and qualifying. As must the thought that, at tracks such as Baku, say, the chance of a red flag is high during the race, which will turn planned strategy on its head and affect tyre availability.

Friday night effectively involves redoing all the work carried out pre-event and running the simulations again – but with a bit more certainty in the model. At the end of it, a Friday night update report will produce the same type of document as the preview, with some – but not all – of the questions answered.

Chapter 18

FREE PRACTICE 3 (FP3)

You would think the final free practice session ought to be approached in much the same way as FP1 the day before. It is a similar routine: travel to the track, have breakfast and prepare for another 60 minutes of running at more or less the same time of day.

But FP3 is different. The feeling of being more urgent, slightly more frantic, begins at the paddock gate on Saturday morning, where a few hundred like-minded team personnel are waiting to be electronically swiped in as soon as the curfew has ended. Once on the other side of the barrier and heading for the garages and engineering offices, you can tell by the brisk pace all around that everyone knows there is important work to be done, not just for FP3, but for the all-important qualifying session later in the day.

Discussion in the van on the way in will have touched on the weather and gone back over the choice of tyres and in which order they should be run, the strategist's first job being to ensure the correct compounds have been switched on in their blankets. Confirmation of the increase in pace and urgency follows in the form of more emails than before and the arrival of data and reports based on overnight work in the simulator and by mission control in the factory. There is much for the strategist to get on top of – after breakfast, of course, although that, too, feels different, more condensed, thanks to everyone turning up at once rather than eating at different times.

When the drivers arrive, there is every likelihood they will have further thoughts on tyres and opinions on the reports and analysis undertaken so far. These fresh thoughts will be along similar lines to other discussions likely to have happened in the various cars and vans bringing the crew to the track. Since such exchanges will have been off-line, the strategist needs to keep across the various ideas and, if necessary, merge them into the plan for the day. It is essential not to miss anything, no matter how trivial or obvious it may seem.

Then, as the pit lane opens for the start of FP3, something strange happens. Everything goes quiet – a rare occurrence at a race track. After the frantic preparation and customary positioning at the pit wall 15 minutes ahead of the session, the green light comes on – and no one goes on track. Unlike FP1 and FP2, there is no need for a test programme or long runs. The number of laps needed is fewer. The day suddenly seems more disjointed when the first 15 minutes of FP3 are represented by a pit lane that has gone strangely silent.

I would spend the time making sure the weather pattern and track temperature were as forecast. This was also an opportunity to check the wind direction based on the flag I had previously picked out as being free from the funnel effect created by the grandstands on the main straight. Given how changeable the wind can be and the effect this can have on the cars in specific corners, this would be discussed with the drivers.

When the cars finally take to the track, the preference is usually for low fuel, less than had been run in FP2. Having made a calm start, FP3 now becomes busy. The focus is on qualifying and the need for the drivers to establish their braking points when the car is as light and as quick as it is ever going to be. Any sensors no longer needed will be taken off the car to reduce weight to the minimum.

While everyone may be on low fuel, the pattern is varied as laps may be aborted or last-minute changes tried before qualifying. It could be that mission control has suggested a different downforce level – which calls for a big change to the car. If so, assuming the job has been done, you will want to go out early and make an assessment. The reduction in downforce increases straight-line speed – which may be good for qualifying but possibly detrimental to the tyres in the race because they will not be loaded as before, meaning the car will slide more in corners and increase degradation.

If a change involves a different rear wing, then an early run is definitely called for, because 20 minutes will be needed to change back to the original specification rear wing if the second choice doesn't do the job as hoped. That's why big steps such as a change of wing or down-force for FP3 will be highly debated beforehand because, whatever happens, the car will be on low fuel, which is not ideal when trying to think about these alterations relative to the race. Rather than making significant changes, the ideal option is running two sets of tyres as planned, quite late in the session.

It could be, however, that your hand has been forced to think about an early run in FP3. If on Friday the track had been wet, or there was a red flag, so urgently needed degradation data is missing, there will be a debate about attempting a long run in FP3. To my mind, that should be a non-starter. I can't recall ever pulling data from a long run in FP3 that was of any use. It was always rubbish because the driver didn't want to know about a long run at this stage. His priority would be to get this out of the way and move on to thinking about qualifying. You would never get a good read on tyre degradation because the driver was always pushing too hard. Someone might get useful readings on fuel consumption or temperatures but, from the strategist's point of view, a long run in FP3 is a largely pointless exercise.

Assuming there have been no distractions and the planned two-tyre runs are taking place, it is preferable not to split the cars to do a test (unless one driver has a specific issue that needs sorting out). The objective is to have both cars as close as possible to what you believe to be the optimum setup.

On the subject of trying a different level of downforce: if the driver who tests the change likes it and wants to continue, the other driver's opportunity to experience the different setup will be limited to his second run. This makes it very difficult to compare things in a fair way, plus the second driver will only have a limited number of laps to try the resulting differences in the braking zone. It will be difficult to readapt before launching into qualifying.

As mentioned, FP3 ought to be about running two sets of soft tyres. It is possible to go for something different across each set. On one, you might try a double cool between two fast laps. On the other set, you might think about a slower out lap. The overall aim is to find the right answer for everything associated with qualifying and ensure the car is set up as close as it ever can be to the optimum for a fast lap.

The driver will be much more focused on qualifying and, with the aid of overlays comparing the best laps of each driver, talking to his engineer about where the car could run even faster. FP3 is also the time to rehearse, once more, getting the car refuelled and turned around quickly in the garage, as well as to practice the out-lap procedure and possibly finish the race preparation that could not be completed on Friday.

A familiar job in FP2 would involve laying rubber with

soft tyres in the pit box to give the driver as much grip as possible when launching from a pit stop during the race. Much depends on the pit-lane surface. Asphalt normally offers reasonably high grip, unlike concrete (which is used in many pit lanes such as Bahrain). But if the FP2 programme had been so hectic that there was no time for a burn-out of rubber, then FP3 is the time to do it. But it's not that simple.

At this stage of the weekend, the actual race is not the priority – qualifying most certainly is. Ideally, you are trying to run as late as possible in FP3 to use the track at its fastest, even though that risks a stoppage – and possible finish of the session – caused by a red flag near the end. With it being essential to have the driver complete his two runs on the soft tyre, the strategist is faced with this balancing act of neither wanting to be too late nor too early in the session. I would have tried to build a picture for this circuit by spending time looking at how many red flags there had been in FP3 in previous years, while also checking the times various teams chose to run.

Part of the overall plan is to have the driver lay rubber when he finishes his first run. But if he makes a mistake on one lap, what usually happens is he gets caught up with an extra lap and forgets to do a burn-out when he returns to the pits. A reminder from the strategist is likely to bring the response: 'Oh, we don't have time for that any more. Let's not worry about it now.' At which point the strategist will say, 'Okay, fine. I'll remind you of that on Sunday when he wheelspins out of the box and loses a couple of tenths!'

In the event of a driver locking up on his first set of tyres in FP3, this is effectively the end of that run. The driver will want to look at the data and check the lost lap against the other driver's in the hope of discovering what went wrong. If he failed to get a lap in, the driver might want to look at the other car's data from this session and compare it with Friday's running to see if any of the braking zones have changed or the car's balance through a particular corner is different. This process of inferred learning from the other car is important before going back on track with the second set.

Drivers are gradually pushing more and more as the weekend goes on. This means they only begin to explore the limit of the braking zones during FP3, the argument being, if you want to discover the limit, this is the time to do it, rather than during qualifying.

The overall picture becomes even clearer at this stage when measuring where your car sits relative to the others in terms of pace. Everyone will have stripped off most of their test items and established fairly clean runs towards the end of FP3. The likely order in qualifying is beginning to take shape.

When FP3 ends, a debriefing procedure similar to Friday is followed, only this time the focus is on what to expect during qualifying based on the lessons learned. The strategist's debrief will focus on the biggest thing that needs to change for qualifying, even though other details might have emerged in connection with the race. There is a two-to-three-hour gap before qualifying. That time goes very quickly indeed and setup decisions need to be agreed, a setup sheet prepared. Unlike the gap between the two sessions on Friday, when fresh setup changes can be incorporated into the session, this time the car must be ready in optimum condition for the first run.

The break between FP3 and qualifying tends to be much more frantic on the race engineering side. The strategist, meanwhile, will be preparing for the session, checking release lines of where your cars enter the track relative to others, and working out timings in terms of how many laps you think you can do. Much depends on predicted pace when establishing how many laps might be necessary to progress through qualifying – and what you believe your cut-off time might be in each of the first two phases. The first revealing test of seemingly non-stop estimates and calculations over the past few days is finally about to happen.

Chapter 19

QUALIFYING

Qualifying has been run in many different formats over the decades but the current system – first introduced in 2006 and tweaked a little along the way – works well. Split into three parts (known as Q1, Q2 and Q3), the hour-long session sees the slowest drivers being eliminated from the first two segments, leaving the fastest ten to fight it out for pole position in Q3.

The key is obviously being fast enough to make it into the top 15 going into Q2 and then, all being well, reaching Q3. Estimating every driver's theoretical lap time (or 'cut-off') needed to progress into Q2, and then Q3, is vital. This is done by looking at the practice sessions, particularly FP3, and estimating the driver's theoretical best time. The lap on every circuit is divided into three sectors with the computer showing the driver's time in each. By taking his best time from each of the three sectors – but not necessarily on the same lap – you arrive at his theoretical best lap time.

There are many corrections to be made along the way: perhaps the driver didn't run a second soft tyre, maybe he made a mistake, or track improvement needs to be considered. The objective is to arrive at the best pace the car and driver are capable of. Taking that FP3 figure and reducing it by the improvement in lap time noted from FP3 to Q1 in previous years, you know what is theoretically possible and therefore what you believe to be the cut-off (or the 15th fastest time) that will be necessary in order to go through to Q2.

During the briefing for qualifying, our expected cut-off time is read out so that everyone is aware of what we think is needed. Assuming we reach Q2, there will then be a new cut-off time necessary to make it to the top-ten shoot-out. The cut-off is an estimate with a margin of, at the most, two-tenths of a second, and is subject to ongoing adjustment depending on whether the session starts off faster or slower than expected.

It is very important to have everyone on the same page. This also applies to the drivers, who could have different attitudes. One could be more confident in his pace than he really ought to be; the other could be the opposite. There are times when two runs are necessary in Q1, but the over-confident driver will complete the first run and return to the pits in the belief that he has done enough. This is when it needs to be pointed out that he was not paying attention during the briefing when it was said that a significant improvement in track condition was expected in Q1. On top of which, a couple of drivers have not had good first laps and are almost certain to improve. Your driver needs to be told: 'We are not safe yet. We definitely need to run again.' Conversely, on another day, the more cautious driver may feel he has to go again and use another set of tyres when, by our estimation of the cut-off, he has already done enough to make it into Q2.

This is typical of a mix of attitudes sometimes found in qualifying. It is necessary for the strategist to know who is risk-averse and who is not. I was probably too risky at times during qualifying; too keen to say I thought the lap time was okay and there was no need to run again. Whereas Tom McCullough, my manager at Aston Martin, was conservative and would opt to run again if there was any doubt. A balance between these two mindsets is best.

The ultimate goal is simply to make it out of Q1 in order to get anywhere in qualifying. But to do that, it is important to ensure everyone is aware of potential weaknesses, such as one driver being able to make do with two sets of tyres while the other might need three. If the latter applies on a circuit in which time allows three runs, it will really push the run plan and squeeze time in the garage to the point where mild panic could be in danger of setting in. Of course, there is an argument that the additional run gives you three bites of the cherry, particularly if you are struggling. The fact is that the strategist has less control with three runs and the time to learn in the garage becomes limited. It is preferable to have two well-controlled runs with time in between to study the data. And there is a lot to think about.

Is it necessary to add extra fuel – and time – to cover a possible abort lap? If so, how will allowing this extra time impact your track position? You are trying to think about where the other cars may be, where you may be, and whether that's positive or negative for your position. We run simulations and go through previous races, all with the aim of trying to get everyone aligned going into qualifying. They need to be aware of how big the track improvement might be, to the point of telling a driver on his out lap about data from runs already completed. A substantial change of five- or sixth-tenths of a second is worth knowing because it will affect performance in the braking zones, be it arriving faster at the corner or having increased grip. It is a case of providing as much information as possible to help the driver be aware of what lies ahead.

Track position is key in qualifying when it comes to finding a gap to allow maximum aero performance. That's not a problem during Q3, but it can be difficult to find space in Q1 when there are 20 cars on track. Even during Q3, there can be crowding at a track such as Monza where straight-line speed is everything and drivers are positioning themselves in the final minute to be close enough on the long straights to take the benefit of a tow from the car in front. In situations like that, you can see it coming, based on what has happened in previous years. On an out lap, if you want to go relatively slowly, then others are going to overtake. In which case, the driver is kept informed about the number of cars coming at him. Often it becomes a badly managed situation by everyone involved and there's not a lot you can do about it. In addition, it's about aligning the requirements of an out-lap track position with tyre performance.

It is comparatively easy to do a reasonably fast out lap and choose your space, particularly if your team has pace in hand and can afford a suboptimal out lap and yet still make it through. Whereas teams struggling to progress might be relying on a perfect out lap and have little choice when it comes to finding a gap. In these cases, everything must align, which comes back to the strategist's heavy emphasis in the briefing on ensuring everyone is reading from the same page.

At Force India and later we had many qualifying sessions in mixed conditions and the reason we might have done poorly was because people did not fully understand what was happening. That would become apparent in the post-qualifying briefing when someone would say, 'I thought the rain was going to stop and the track would become drier,' while someone else would say they thought it was going to become wetter. Apart from this being another demonstration of the mix of optimistic and pessimistic personalities, it heavily underlines the need to have everyone listen and take on board the weather prediction – usually accurate – given by the strategist.

Weather, for example, came into play during qualifying

for the 2023 Brazilian Grand Prix. Black, menacing clouds were gathering on the horizon just before the start of Q3. With Brazil's reputation for sudden and heavy downpours, it was a case of having both drivers ready to go as soon as the track opened. But that wasn't the case for some teams and they were caught out. The briefing should have stressed the need, in the event of threatening clouds, to have everyone on the ball; the first lap is likely to be the one that counts and everything needs to be perfect. The arrival of rain will ensure there is no second chance.

There must be no doubt about this, particularly when, as was the case at Interlagos, one radio message said it would be seven minutes before the rain started. That may have been true for the main bulk of the rain, but the fact remained that the light sprinkles preceding it were enough to affect a track surface that quickly became greasy. The first lap out was critical. It is a point that cannot be stressed enough, because the optimistic team members might have been thinking, because the heavy rain had not happened immediately, it might not come at all and luck was with them; everything was going to be all right when, clearly, it wasn't.

Which leads to another important point. The driver needs to be reminded that, no matter how poor he feels his lap may be, this is IT. Do not sacrifice that lap in the belief that the next one will be better. Push on to the end. Everyone is in the same boat. If there is a question over the chance of squeezing in a second lap before the rain, it is possible to run an engine mode that allows a second consecutive lap by using 50 per cent of the battery rather than 100 per cent for one lap. Once again, efficient communication and understanding among team members is essential for that to work.

The Brazil incident illustrated how difficult it is to be fully on top of everything. While race fans watching from home could see massive black clouds filling the top half of their TV screens, strategists and others were buried in the data coming across their monitors. There should be at least one person on the pit wall taking the time to come away from the screens, take a quick look around and inform everyone (particularly at mission control), 'The wind has picked up.' Or, 'It's suddenly got darker.' Even then, there is always a TV running somewhere on the pit wall or in the garage showing the race on the world feed. It doesn't take much to make a simple comment on the intercom to say that the commentators have noted umbrellas going up in the grandstands at the far end of the circuit. The worst feeling is to know that 18 cars have left the pit lane and the two remaining in the garage are yours. It's a sure sign that it's too late; the ship has sailed. It can be difficult because the crew has been sitting in the garage watching the TV screen, noting that everyone has taken to the track and wondering why their cars are still sitting in front of them.

The rain in Brazil in 2023 highlighted the argument that

F1 teams ought to have someone positioned somewhere other than in the pit lane. At Interlagos, for example, the back of the paddock gives a great view over the rest of the track, with São Paulo in the distance. You can see for miles. The problem here is that unless that person does this at every race, it is difficult for the strategist to judge the observer's perception of what's good and bad, because that assessment may not be the same as yours. Some teams place people around the track and give them a phone to report back about the rain being at level 1, 2 or 3: whatever had been agreed beforehand. Such information would probably be too late for qualifying, although it could be useful during the race.

When I was at Aston Martin there was a suggestion at one race that Lawrence Stroll's helicopter should be sent up to report on the changeable weather. I was not in favour, my argument being that we had enough people involved and I didn't need another source of information, particularly from someone who had never spoken to me about rain before and I had no concept of what their understanding might be. I politely declined.

Compared to practice sessions, qualifying becomes calmer the further you go. There are fewer cars on track; less time, perhaps (qualifying reducing from 18 minutes in Q1 to 12 minutes in Q3), but more space. Out laps are more clearly defined; the same for in laps. There is a better understanding of where you stand compared to your closest rivals and their tyre choice. Having made it out of Q1, the pressure is reducing, as is the choice of tyres. In Q1 there are many questions; by the time you reach Q3 it is a matter, based on the number of tyres left in your allocation, of using a new or used soft tyre. The cut-off will become more sharply defined; the overall picture becomes easier to understand. In the event of a red flag, there is slightly less drama than you might expect. In qualifying, the clock stops, whereas during free practice it keeps running and sets in motion the questions and guesswork outlined in previous chapters.

One of the biggest issues for a strategist during qualifying is knowing when to talk. Use of the intercom is best summarised by 'right information, right time'. Depending on where you are in the world, an out lap can be frantic as you try to get across to the driver his position on the track and how much time he has available to get to the line and start a lap before the chequered flag appears. He will also need tyre temperature and engine information. There may be new advice on how he might improve his lap time, based on his previous lap or perhaps a setup change in terms of a brake balance or differential setting.

In the middle of this, the strategist is looking for a suitable gap in the traffic while allowing for enough pace – be it fast or slow – to have the tyres reach the right temperature and make sure the battery is fully charged. That's why the most difficult aspect for a new strategist is knowing when their traffic information is more crucial than tyre, engine or any other data being given. In which case, you need to interrupt whoever is speaking at the time. It is about knowing when your strategy information is a 'must' rather than a 'nice to have'. At such times, when everyone wanted their say, I always felt my previous experience in performance engineering and some of the other roles helped me understand what was important and why.

I would try to pass on information to the race engineer before the driver has crossed the line to finish his lap, then allow the driver and his race engineer to do their performance communication before providing more input as needed. It is the race engineer who is the link to the driver, passing on details received from strategy, as well as the engine and performance engineers. Those are the main sources feeding the relevant facts to ensure the car is at its optimum for the beginning of the lap. If one of those small but important pieces fails to get through, the tyres may not be the right temperature, or the engine will not be in the correct mode/setting, or the battery not fully charged.

Taking the battery as an example, this would be charged under normal circumstances by pushing with full throttle on the straight. But in qualifying, it is often the case that the driver is also trying to let others overtake to create a gap on track, in which case the battery might not be fully charged. It is often a compromise between ticking all the performance boxes and establishing the best track position.

As I said, Q3 is easier because there is less traffic to manage and the gaps are more readily available. It also helps that by this point the driver will have had two sessions to become acclimatised. Or, at least, that's the theory.

With the pressure on, the driver could push too hard, lock up and flat-spot a tyre going into the first corner, sending all of this carefully planned preparation down the escape road. Watching this from the pit wall, what do you do? My attitude was never to sigh and think about what might have been. There is no point in wasting time and energy on the emotion of what you've just seen on the TV monitor. That opportunity has gone. Get over it. Thoughts should be about where we go from here. Is that tyre recoverable? Can we abort and try again on the next lap? Your mindset very quickly changes to what might be possible.

It is more difficult for the strategist if the mistake comes from their side: for example, a missed pit stop or safety car in the race. That would take me much longer to move to the position of 'forget it; move on' and helped me realise that the driver having, say, made his error on the first lap of Q3, would probably be feeling the same. Nonetheless, I knew I could have no influence on how the driver might be dealing with the setback.

Once qualifying has ended, the race engineer will talk everything through with the driver, while the strategist will quietly wonder if any part of what happened at the first corner, say, is going to land on their plate. Is there potential for the fallout coming back to you? Could it be that you ran at the wrong time? Did you have a different track position compared to everyone else? Were the wind conditions different at that particular moment? Was there a contributing factor that could prompt someone to say, 'We shouldn't have run at that time'? Or, 'We shouldn't have run that tyre'? Or, 'The tyre temperatures weren't quite right'? The strategist would quietly check everything and be prepared to deal with any questions that might arise.

The most immediate and widespread action, post-qualifying, is to check the video, particularly if a suspected penalty is on the cards. That's why the strategist will ensure that someone in mission control makes the video available when those who need it come in to find an explanation. It is good to be prepared for the questions. It is not about justifying your position; more a case of simply explaining what went on and the reasons behind your thinking.

F1, as a whole, is strong and pragmatic when it comes to learning lessons. If there is something that should have been done differently – but wasn't for understandable reasons – that will be taken onboard and understood. People are honest enough in their analysis to say, with hindsight, they would have acted differently, and so will correct or find a solution going forwards. Fine. Let's get on with talking and preparing for the race now that qualifying is over.

Chapter 20

POST-QUALIFYING

If you finish qualifying in 11th place (P11), they call it 'Bolters' Pole' – a reference to not making it to Q3 and allowing the mechanics (affectionately known as 'bolters') to make an early start on race-day preparation. Despite the pit lane now being under Parc Fermé rules, there is still a lot to be done.

Parc Fermé was introduced to stop teams building a special, lightweight car for qualifying and then racing a totally different specification more suitable for the Grand Prix. Parc Fermé comes into force the minute qualifying begins and runs until the start of the race on the following day. In simple terms, you race the car you qualify. That makes it tricky to find a setup that will be good for the race but not impede performance too much for the very different demands of a fast qualifying lap. Or it could be the other way round, depending on the circuit – Monaco, for example, where qualifying takes priority.

The difficulty from a strategy point of view is making your best guess at what the temperatures might be in the race and how that could affect the car's cooling specification. Much depends on where you think you might qualify. If you are confident of running at the front, or find yourself in the happy position of knowing a start from pole is likely, the question becomes: do you make the usual allowance for the car running in traffic? Or do you assume you are always going to run in clear air? The difference between how the car is set up for each can be significant in terms of pace. But if you do not allow for traffic and then fail to win pole, it's going to be a difficult race in terms of engine temperature management.

The Parc Fermé rules say that if the ambient temperature shifts by more than ten degrees between qualifying and the race, a change of cooling package is allowed. But the reality is that ten degrees is a big swing and rarely happens. The strategist is left with the very difficult job of making an educated guess on Saturday about the likely ambient and track temperature 24 hours later, cloud cover being the most difficult thing to predict. You want a cooling package that is efficient and fast for qualifying, and yet allows you to race well too. Changes are allowed if the race is wet and qualifying has been run in the dry – or vice-versa.

My job was to foresee the temperatures by using up-to-date weather reports. The engineers, meanwhile, would be trying to establish the best cooling level for the car. Once that is settled, the setup on the car leaving the garage to start qualifying will be the same for the race. Changes to suspension or downforce are not allowed once qualifying has commenced, although, obviously, tyres are changed and fuel added, along with the driver's drink supply.

Yet despite all the above, it is not widely appreciated how much of the car comes apart during preparation for the race once qualifying has finished. It's not as if the car arrives back from qualifying and then sits all buttoned up until it's time to race. A substantial amount of the bodywork is removed to allow checks to be made in many areas. The rear corners will be taken apart to inspect the brakes. Items will be marked with little stickers to ensure the same part goes back on. All work carried out on the car is watched closely by an FIA scrutineer. A camera is running from the moment the car, along with any parts yet to be refitted, goes under covers for the night.

The clutch position to be used at the start of the race will have been set before Parc Fermé comes into force. The effect is similar to slowly letting out the clutch on your road car and finding the bite point. But rather than using his left foot, the driver has two levers which he pulls. He holds one all the way to fully engage the clutch. The second lever is held at the bite point, which means he can drop the other one totally and hold the second, the launch position being set electronically.

The exact point for the best launch will change a little, depending on which tyre is chosen. That's why, before qualifying, the strategist has been taking a stab at working out the most likely start tyre. It is usually a soft or medium. The hard tyre is unlikely to be chosen – unless things have gone badly wrong during qualifying and a start from the back of the grid is inevitable. In which case, the hard tyre will be used for a long first stint in the hope of making up positions when others make their first pit stops. But, arguably, if you're at the back, the launch is not particularly important in any case.

With all the settings – be they car cooling, suspension, downforce level – having been decided, the emphasis shifts totally to strategy. You would be surprised at the number of strategists suddenly appearing on Saturday night! Everyone has something to add to the conversation about race tactics, even though many of the points will have been raised earlier in the weekend – or, indeed, during the pre-event discussions at the factory. The strategist, now that grid positions have been established, will have begun to look in greater depth at the models that have been running all weekend. There is likely to be a different take on everything now that knowledge of rivals' tyre choices is available, along with grid positions.

As happened after the first day of practice, many of the pre-event questions can be answered. The model is becoming tighter and, hopefully, more accurate. You will have a fuller knowledge of likely tyre degradation, even though it covers a wide range. In which case, on Saturday night, the strategist will run a model for higher degradation and one with lower deg, to cover all possibilities.

This may or may not have been done in time for the first strategy meeting (there having been a debrief in the afternoon, which focused on events during qualifying). Saturday night is often when the strategy model is questioned, a typical comment being, 'Oh, I don't think pace/deg will be that bad.' In simple terms, the debate will range between people thinking the car will be stronger than predicted, or not. Some of the speculation could be based on the team having had a better qualifying than expected and may lead to suggestions that the car is actually much more competitive than the strategist's apparent pessimistic projections. At such a point the strategist will go back through qualifying and say, 'Well, these three drivers [who we expected to be ahead of us] all made mistakes. If all three had done what we think they are capable of, we wouldn't have qualified as high as P6 or P7.' This conversation can be very difficult at the start of the season when there is not a good basis for accurately predicting the car's pace, or how the race might pan out. Life becomes problematic for the strategist when you start, say, sixth on the grid and finish outside the points. It can feel very negative.

By Saturday evening the strategist is often torn between accepting that, obviously, we are going to be the very best we can be but, realistically, we're likely to finish 11th rather than maintaining our position further up the starting grid. Alternatively, you could project a finish in the top six, finish 11th and thus create the impression that the strategy was wrong. It is very difficult to ground expectation in realism while trying to maintain some optimism. As soon as you start to show graphs of where you think we are and suggest plans A, B or C, it's not long before feedback comes in suggesting, 'I think we're better than that.' Or, 'I think we're worse than that.' Even, 'I think Williams are worse than you say.' And on it goes. It's not surprising that those meetings can be quite frustrating for the strategist.

Drivers and race engineers will give some input at this stage; systems engineers responsible for the start performance will have their say; the tyre engineer will give his thoughts. The strategist will be aiming to balance the various views. You are trying to run through a lot of scenarios and come up with the potential opportunities and the possible pitfalls, much of it based on what has happened in races from previous years.

Discussion on Saturday night is still very much a discussion. It is not necessarily the final decision on strategy. The strategist might have a feel for what ought to be done, but there will be a lot of debate as the evening wears on. To present some solid facts and answer doubts about the tyre model or car pace or whatever is causing the most disagreement, the strategist might run a sweep of the projected suggestions and then say, 'Look, this is our preferred strategy, but this is what happens if we run it with the high-deg model. And this is what happens if we do it with low deg.' You work quite closely with the engineers in the factory because they will be running some of the simulations for you and helping to investigate similar models.

Although the cars must go under covers, there isn't a curfew in terms of when you need to physically leave the track on Saturday evening. Things have improved since my early days at Force India when you returned to your hotel and either ate in your room or had a McDonald's or something similar. Trying to work on a laptop in your room under those circumstances really was suboptimal. In later years, we would have dinner at the circuit on Saturday night. That meant good food and the chance to continue working in your office situation while still fully connected to the internet – which could never be guaranteed in many hotels, given their poor WIFI.

Working at the track was always preferable thanks to being able to feel more together with the engineers in the factory, who would continue preparing reports and working long after I finally left the track. Since I often drove one of the vans, those travelling with me would either busy themselves with analysis or go for a run around the track while waiting for me to finish. That's why any race where we could walk between hotel and track was much better, because there was no feeling of holding anyone back.

By Saturday night the strategist has reached the stage of looking for things that may have been missed earlier. You might go back through the tyre model arrived at on Friday night and check if one of the outliers discussed has sneaked through. Is there anything coming out of your result that doesn't necessarily ring true? Is it saying start on a hard tyre and run the whole race on that? Do you believe the answers that you're getting out of the simulations? Everyone in the engineering office is thinking the same way, questioning everything. With so much data flowing back and forth, it is very easy to overlook something or make a simple but potentially critical error.

Having spent the previous few hours trying to communicate with a lot of people, I used to enjoy the change of pace as the office became comparatively quiet on Saturday night. It would remain busy into early Saturday evening, as various people drifted across and began asking questions just as I was trying to write my reports. One by one, they would leave. It would come down to perhaps just three or four of us. Not only was there space to spread out, but it was also the perfect environment to crack on with the strategy. Then it would become really quiet in the pit lane and paddock. When you thought about how frantic the paddock and pit lane had been when under full steam earlier in the day, the same place in the darkness, with hardly a soul to be seen, felt eerie. There would be one security guy at the back of the garage but no one in hospitality when you went over to make a cup of tea. By late evening, there was little else to be done. Most big decisions had been made, but important choices remained. They were for the morning – race day, the climax of the weekend.

Chapter 21

SUNDAY MORNING; GETTING READY

Wake up on Sunday morning and you feel this is 'Game On'. It's race day, the bit that you're here for, the part of the weekend when you can make a difference. It's something I felt when a strategist, and I feel it now when working for Sky. The previous two days have been about preparation, speculation and taking steps from which, if wrong, you can recover. On race day, this is it.

The day gets off to a strange start. If it's a European race (and for some flyaways), you will have checked out of the hotel, ready to fly home that night. There is a knack to packing your bag in a way that ensures your travel kit is easily accessible. It needs to be somewhere convenient because, quite often, you will be leaving the track and heading for the airport very soon after the race has finished. That may not be a problem for the lads as they change into their travel gear in the car park, but it's obviously a different story for female members of the team. I would prepare a small separate bag of whatever I needed for the homeward journey and bring that with me into the track, having worked out somewhere I could get changed. Having to think about this in advance felt like a bit of a faff that I really didn't need to be dealing with first thing on race morning.

The strange feeling continues on arrival at the track. Although there is not a curfew in place for team personnel, there is a set time for the removal of covers from the cars. The period in between is spent having breakfast, which seems comparatively relaxed, particularly when there is usually about five hours to go before the start of the race.

The strategist, meanwhile, is busy. On a Friday or a Saturday morning, others on the team are much more important: the race engineers, for example, as they work on car setup. On Sunday, however, it seems that people who have ignored the strategist all weekend suddenly develop an interest in what's happening on that side! Questions will cover anything from pit stops and tyres to your thoughts on any reports received overnight from mission control.

The weather is usually a constant topic. The strategist will have already checked if, say, it is as cloudy as anticipated. The temperature can often be the cause of uncertainty among those not realising that we are operating at a slightly different time on Sunday morning. Frequent comments such as, 'Oh, it's cooler than yesterday' – or, 'It's a bit hotter. What d'you think?' – need the reassurance that everything is okay because we are an hour later, or earlier, as the case may be.

Any significant discussion in the van *en route* to the track will need to be repeated to the guys on the other side of the garage. The tyre engineer might pass on something he's thought about overnight; someone else might ask if we have considered an incident that affected a previous race at this track. There can be any number of last minute thoughts which may be thrown in randomly but are always worth consideration.

Around mid-morning, most teams will have their final strategy meetings with the drivers and engineers. The overnight reports will be discussed but, inevitably, the main subject will be tyre selection for the start of the race, particularly if the choice is not obvious. Given that the first person any driver wants to beat is his teammate, part of the problem becomes each driver looking to do something different.

Depending on a driver's personality, and where his head may be at the time, it can be difficult to convince him of the merits of what the strategist considers to be the best start tyre. One driver could prefer the soft tyre, because he feels his advantage lies in making a better start. In many ways, this is about perception based on what each driver thinks is better or worse, and which option they can work to their advantage. The various arguments will go to and fro as discussion covers the examples looked at and the simulations we have done when trying to highlight the potential benefits and negatives of each strategy.

These conversations can take quite some time and, once the meeting has finished, they can continue offline between drivers and their engineers – in which case the strategist will do their best to be involved in both discussions, taking care not to prioritise one driver over the other, and avoiding going to same driver first each time. I tended to choose the driver for whom I thought the tyre decision was the most difficult. The decision can be fairly clear cut for the driver that's five or six positions ahead of the other. It's often not so easy for the driver farther back. Equally, the strategist needs to prioritise the driver with the best potential for scoring points. It can be very difficult to get that balance right.

Some discussions would be longer than others when it came to reaching a conclusion. The conversation might trickle on because, having decided on one thing, the driver could be on his way to hospitality when he thinks of something else and returns to make the discussion feel like it is going on for ever. That didn't happen often; generally, we'd get there in the end without too much debate. Meanwhile, the strategist would be checking either for news of rivals making changes to their cars, or for grid penalties affecting other drivers and, therefore, your grid positions.

You would discuss how easy it might be for us to overtake X, Y or Z, depending on how our straight-line speed had looked during qualifying. It's a case of trying to build a picture of which way the race may go while highlighting your strengths as well as potential difficulties. We usually set out three plans which could be designated A, B and C. It wasn't always a case of choosing letters in alphabetical order; some people might prefer S to align with a safety car plan, for example.

Having both drivers on the same strategy makes it reasonably easy to establish a plan, as is the case if the strategies are very different. But having the drivers on strategies that vary only slightly can be a problem. Do you, for example, choose target stop laps that are the same for both drivers?

I tended to pick stop laps that would be easy to remember. If, for example, the optimum stop lap was lap 18, I would make the planned stop for Plan A, lap 20 because 20 is easier to remember. We would then work in Plan A, target minus two. Using round numbers can be helpful to the driver, who has a lot going on.

There would often be a final meeting to outline the overall plan for the day. With strategy having been discussed in detail at the earlier meeting, I would use this occasion to quickly list the proposed strategies and the expected safety car windows (which I will explain in more detail in the next chapter).

It is important at this stage to make anyone not involved in strategy – for example, the engine guys – aware of the planned pit stops. If the car is due to come in on lap 18 and the engineers can see the engine is running hot, they need to let everyone know. Steps must be taken to have the driver back off slightly or run in clear air beforehand to prevent the engine from overheating when he comes to a standstill – no matter how brief – in the pits. The strategist will be aware that having their driver run too close to the car in front could be detrimental to engine cooling.

This is why the strategist's continual need to have everyone working to the same plan is so important. We need to know the intended fuel saving; we need to be aligned on brake saving that may be necessary if the circuit has particularly heavy braking zones. Now is the time to ensure full awareness of the start tyre, stop laps, traffic and back-up plans.

Passing on information about the intended pit stops is an interesting subject. The pit crew will be ready to spring into action on every lap, but it is good that they know when the planned stops will be, and how many. Some teams prefer to keep that information from the pit crew beforehand for fear of someone telling his mate in the garage next door. That's how cautious you can be when keeping planned strategy closely guarded. In view of the need to keep everyone fully informed, I always felt such secrecy was a strange thing to do within your own team. Everyone is there to try to win!

As the last check of the cars takes place in the garage, the fuel is added, the final figure being influenced by the number of planned stops. A pace managed one-stop strategy will use less fuel than a pushing two stops. Equally, you know consumption will be affected if your drivers look likely to spend a lot of time in traffic on a circuit where overtaking is difficult. Or it might be worth checking how detrimental fuel saving could be when running on worn tyres in the final ten laps. Fuel saving on worn tyres will drop the tyre temperature quickly and can lose lots of performance. It is possible to ensure carefully planned fuel saving does not completely destroy your race. The end game, of course, is the weight-saving that comes with carrying less fuel at the start. All these considerations will have been part of the race-morning discussions.

At some point, I would go to the pit wall and change the computer setup, which would be different for the race compared to practice and qualifying. Apart from using special graphs, the entire software had to be set up to cope with procedures unique to the race. Quite a few things needed adjusting.

The final pit-stop practice will have the crew fully kitted out in their helmets, exactly as per the race, before checking that all equipment is primed and ready. Having tried to watch as much pit-stop practice as I could during the previous two days, I would make a point of being there for the final run through on Sunday morning. It's surprising how much confidence – or not – this can give going into the race when deciding whether to be aggressive with your strategy. In the event of a safety car, say, it could be that both cars (if they are close in the race order) will need to come into the pits together, with the second car waiting – or 'stacking' – behind the first. That's a big call for the pit crew, and the strategist might think twice about stacking the cars if pit-stop practice had not gone well on Sunday morning. Mistakes are rare during final pit practice as, being race day, everyone is totally focused and sharp.

After the final pit practice comes a lull in the pit lane. The drivers go off for a parade of the track, either as passengers in classic cars or on the back of a flatbed truck. With between 30 and 45 minutes to go before taking position on the pit wall, it is a good moment to get away from the rising tension in the office, have a quiet sanity check, sort myself out and find time for lunch. I don't operate well without food, particularly when there is a grand prix about to get under way!

Knowing that once the race starts, there's no going back, I would carefully make sure I had gathered everything together before going to the pit wall. From my time with McLaren, I liked to have a paper back-up of quite a few things, such as a very simple sheet on which to tick off the laps, or the race plans A, B and C written down. I didn't like to rely on having vital pieces of information solely on the computer.

Following the usual procedure of walking to the pit wall 15 minutes before the pit lane opens and cars leave the garage, it feels a long time before there is any serious action. During the 40 minutes before the start, the cars complete their laps to the grid. This can be a busy few minutes for the strategists. You want to do a practice launch at the end of the pit lane, which means ensuring some space and time for that. The drivers might then want a gap to allow a feel for the car in clear air; perhaps they might want the opposite, to get a sense of the car in traffic. There needs to be space for the drivers to push.

The strategist is keeping an eye on the clock, ensuring the cars leave the pit lane in time to allow, if necessary, a return for a change of tyres or front-wing angle before the pit exit is closed. Once the cars are on the grid, the situation is outside the strategist's control. There might be a last-minute discussion on strategy or tyres, but that is rare. Normally, the race engineers are in charge at this late stage.

The one exception is a wet – or potentially wet – race, which can lead to a hectic moment or two as decisions are reached about a wet or a dry tyre. Or, perhaps, a change to strategy by choosing to run a longer first stint than planned because approaching rain can be seen on the radar, and it will be important to stay out as long as possible on the dry tyre before changing to wets at the right moment.

Although I liked having access to the data on the pit wall, at times I would feel very disconnected from the grid, a long way from what was happening on the ground. On the other hand, the exposed position on the pit wall afforded a better view of the weather. So close to the start of the race, the grid would be crowded with personalities, media and TV crews as the mechanics carried out final checks within cordoned-off areas around each car. Despite sensing the buzz of anticipation on the grid, I preferred to stay on the pit wall. Listening to the commentators on my TV feed, I might pick up useful information relative to another competitor. The weather radar would provide the latest on-track temperature, wind and other vital information. Meanwhile, I would be in full contact with mission control. The team would be monitoring radios and perhaps gathering useful information from conversations with engineers as the drivers settled into the cars. With all of this going on, I never felt being on the grid added anything useful.

As was proved in Hungary 2022, on the one occasion I left the pit wall. Being my last race before leaving Aston Martin, the team allowed me to go to the grid – rather like a leaving gift. Despite such very good intentions, it was a stressful experience. I had my photo taken with the mechanics by the cars. I do enjoy the atmosphere on the grid – but not when working as a strategist on an afternoon when it started to spit with rain. Had the sun been shining, it would have been lovely. But I was panicking because I couldn't see the weather radar and needed to check the data. It was a very thoughtful gesture by the team, but I didn't enjoy it because I kept thinking, 'I really don't need to be here. I've got a lot to do on the pit wall.' I left the grid early to get back to my preferred pit-wall position and watch the radar.

With the hive of activity on the grid, it feels eerily quiet on the pit wall. Usually there would be just me and perhaps Andy Stevenson, the sporting director at Aston Martin. With about three minutes to go, a significant moment occurs when information begins to come through about the tyres rivals have chosen for the start. The data from the different teams pops up on the screen in quick succession. This is a moment of truth. Are we aligned? Is our choice completely different to everyone else's? Is this as we expected? Or has someone done something completely random? If you're on soft and everyone else pops up on medium, it's, 'Aw . . . shit! We've got this wrong. We're totally out of bed with everyone else.' It's rarely, 'Yay! We've master-strategised this!'

While quickly digesting the information, you are passing it on to your drivers via their respective race engineers. The next move is to update all your software predictions with the tyre details to ensure the programmes kick off in the right way. While that's going on, questions are coming in: 'What do these tyre choices mean? How are we looking?' Quite often, there is mild panic in the voice.

Meanwhile, engines have been fired up and everyone is waiting for the final seconds to tick by before the cars are released on the formation lap. It seems to take for ever. The previous sense of calm has long gone, only to be replaced at this point by a feeling of helplessness. The strategist can do no more – for the time being. The serious action is about to start.

Chapter 22

WE'RE OFF!

The start of a grand prix. It's a massive moment, one that everyone has been working towards for days. You are about to discover if all the careful planning, preparation and speculation is going to pay off. Pulse rates are soaring; everyone is tense. Twenty cars are unleashed in an incredible blast of sound.

There is an undeniable feeling of lack of control as the entire field is as closely bunched as it ever will be during the weekend. The strategist's simulation will have been based on grid positions; say P5 and P6. But, by the first corner, your drivers could be anywhere. Full focus is needed when determining the latest positions and checking that your guys did not overtake anyone off track. If they did, it needs to be established how many places must be given back before a penalty is handed down by the race stewards.

Everything is being monitored in some way: from the extent of damage to one of the cars in the event of contact somewhere on the first lap, to the individual drivers, who are watched by designated team members. The process has been helped in recent years by improved software that allows replays of the GPS or any incidents. A sense of control begins to return.

As part of the pre-race planning, we would have chosen what is known as a 'bail-out tyre'. This would be automatically fitted in the event of a driver coming into the pits following an incident on the first lap. The bail-out would always be ready, the theory being that, even if nothing is said by the strategist, that tyre will be fitted by the pit crew who would also make a matching adjustment to the front-wing angle to allow for the different tyre compound if necessary.

One thing that might also prompt an early stop could be the realisation that, because of tyre choice, the car directly in front is going for a long first stint. The immediate question is: should we bring our car in, change tyres and try to find clear air? When it comes to tyre choice further into the race, there will be a discussion every three or four laps about which tyre should go on next. That tyre will then be brought to the front of the garage in readiness for the stop – whenever that might be.

In the meantime, each engineer will be asking his driver about the car's balance and the front-wing adjustment – if any – he would like at the next stop. This ongoing process means there is a lot happening despite the outward impression that the cars are going round and round and everyone else is killing time waiting for the first pit stop.

The assumption so far is that, if your driver started from, say, sixth on the grid, he will be in that position at the end of the first lap. But if, for whatever reason, he has dropped to somewhere like tenth or 12th, the software will have automatically updated the race plan. Meanwhile, the strategist is reviewing the situation and adapting. If the drop down the field into traffic has affected the original plan for a long first stint in comparatively free air, the strategist will consider an early stop, ask what that would mean in terms of running in clear air, and how that will affect the race for the team. The software will be doing its best to update and, meanwhile, I will have got into discussion with mission control.

The conversation might start with me saying, 'I think we should consider going long because . . .' and I give my reasons. Mission control will then create a new plan showing how they think that race will go. Or it could be the other way round, as mission control makes a suggestion. Either way, we will share our various plans and I will aim to justify whatever decision is reached once everything has been considered.

If, say, positions have been lost on lap one, then the ramifications of a possible safety car come into play and the possibility of losing too much ground because you are so far down the field. In effect, you are rethinking what is known as 'the safety car window': calculating the pit window for a tyre stop under safety car conditions. The key factor in both cases is working out where the driver will rejoin the running order after the stop. Will he be in traffic and perhaps stuck behind a slower car which you know to be running a long stint? If so, the benefit of the stop for fresh tyres may be negated.

The reference to the safety car window 'opening' works like this. If, say, we are planning to stop on lap 20, then our safety car window may open a few laps earlier on on lap 15. This means if a safety car appears on or after lap 15, we make the pit stop immediately thus saving time as the rest of the field is forced to circulate slowly. But, then again, the safety car window can be affected by many things: not only track position, but the choice of tyres available; and if this is the final stop, then the number of laps remaining could be critical when related to the life of the tyres about to be fitted. It is not a straightforward decision but, once it has been made and the strategist says the safety car window is open then everyone – including the driver – is ready for a pit stop from the moment the safety car is called.

The pit wall makes the safety car window call. The only time in my experience when the driver alone made the call came at Monaco in 2014 when I was working with McLaren. The pit loss time, because of a lower pit-lane speed limit (60kph), is considerable at Monaco. Jenson Button, coming towards the end of a lap when the safety car was called, knew he was in a safety car window and made the quick decision to come in immediately. Jenson called it correctly.

Of course, when planning a pit stop, you don't always know what others might do. Sometimes that doesn't matter because a particular pit-stop call will be the quickest for your race, regardless of what anyone else may do. On other occasions, it will matter because you're trying to justify how many positions you will gain or lose at that point in the race. You might be saying, 'We'll have fresher tyres, but we'll lose three positions.' You establish the best- and worst-case scenarios. One might be: if you stay out but everybody else makes a stop, what does that look like? Or, if you box and everyone else stays out, how does that affect your race plan? Those would be the extreme – or 'edge' – cases. If we're not far off the end of the race, you might say to the driver, 'If we stop, you will lose four positions, but you would have a new soft tyre. How do you feel about that?' He might reply, 'These tyres are going to be terrible if we continue, so let's stop.'

You are also taking into consideration tyre degradation and whether overtaking is easy on this particular circuit. There is also the question of the number of tyres available at this stage in the race. Plus, you have to weigh up who you would lose position to. Things have changed since I was with Aston Martin but, at the time, losing position to a McLaren would have been a very different proposition to finding yourself behind a Williams, because the McLaren would have been more difficult to overtake. There is also the ever-present risk that you could have a slow stop.

All these factors need to be evaluated – often, quite quickly. There would be a good debate on the pit wall; an extension of our regular discussions about what to do in the event of a safety car. Even during a one-stop race, we will make hundreds of decisions in the hour and a half, or however long the grand prix takes.

Assuming the race runs to plan, a lot of preparation is necessary for the scheduled pit stop. You have a target lap but, in the meantime, you are trying to judge the reactions of others, particularly the car behind you. Will they try to stop early in order to perform the undercut by using the fresh tyres to go much faster than your car is managing on worn tyres and so be ahead when you emerge from the pits after your stop? Do you think your rival has the pace to do that? Should you try to undercut the car ahead, even though that may be outside what you would expect to do? Or should you stick to the optimum? And is your estimated optimum still correct at this stage of the race?

Second-guessing is common. Is a rival going to be more aggressive than you? Is your stop lap likely to be either too late or too early because of what someone else is going to do? You are continuously re-evaluating the fastest strategy from your position at that particular point in the race, while taking in the reactions of others and any updates to your model that have come to hand.

It is very difficult, particularly when the traffic is heavy during the first five laps, to get a read on either the pace or tyre degradation. It takes about ten laps before you begin to get a good feel for how the race is going. But the picture can be complicated by events and incidents before the opening ten laps are completed. Regardless of the situation, you are checking the information that you already have relative to your model, be it tyre degradation, pace of the car, or track improvement.

It is a case of continually justifying your model throughout the race. Are your assumptions right or wrong? Do you still think it's a two-stop race? Does the hard tyre continue to be the best option for the middle stint of the race? Is there evidence from another driver already running the hard that this tyre will not be as good as we had hoped? Does it appear to have a slow warm-up lap immediately after the pit stop? A driver, starting from the back of the grid on the hard tyre may look slow. Is this because of the tyre, or does the car have damage from a first-lap incident that we don't know about?

You are trying to build a picture, but asking lots of questions about why the picture might be wrong and, at the same time, keeping the driver informed (through his race engineer) about the pit-stop plan. 'We're still on Plan A.' Or it might be 'Plan A, minus two', as in the previous example.

The communication can become a two-way street if the driver feels he wants to stop sooner than planned, but the strategist knows that is not the best option because of something the driver cannot see. He might say that the tyres are terrible. Or he could say the tyres are really good. In each case, the strategist is quantifying these comments against the lap times of others. It could be that what the driver feels to be a good or a bad lap time is actually being replicated by others due to track conditions. The tyre may feel bad to your driver, but the advice will be to stick with it because everyone else is on similar lap times. A lot of it is perception. I found that drivers can be very good at telling the difference between track improvement and tyre degradation.

It is possible to have a significant track improvement which is offset by measuring big degradation, meaning the lap times are very flat. This is something else that needs to be taken into account when working out when the next stop should be. If someone else boxes ahead of you and it immediately becomes apparent how quick they are on the new tyre, you need to be on the case, telling the driver about what has happened while trying to work out the difference between deg and track improvement.

All calculations can be affected if, say, a front-wing end plate on your car is damaged during a collision early in the race. Is the damage bad enough to compromise performance or cause bits of bodywork to break off? The team will have previously gone through a proving cycle with the FIA to show a damaged wing could, in certain conditions, remain intact rather than disintegrate. If the stewards believe a failure is likely, they could force the car to stop.

Performance loss in the event of damage is measured by fine aero tubes on the car, allowing the engineers to establish the loss in load, or downforce. Is this loss different from one side of the car to the other? Could a front-wing adjustment on one side during the next pit stop help return the car's balance? The question for a strategist becomes does any lap time loss outweigh the time loss by changing the front wing in the next pit stop or should a small loss every lap be accepted to avoid the need for a slower pitstop?

The planned pit stop is obviously a key event – but it is another of those moments when the strategist has no control once the stop has been called. You will have called in the driver at a particular point because of a fight with someone for position, and you want to either come out in front or, having stopped ahead of your rival, get there by performing an undercut. Once the process is under way, the strategist can do nothing further but watch the driver come in and clear the stop. Then it's straight back to your computer to see where he's going to emerge on track.

Even if you choose not to turn around, the strategist can tell whether the stop is good or bad just by listening. Having seen and heard pit stops so often, your senses become tuned to every sound. At which point you will tell mission control the stop was 'good' or 'bad' and the reason why. A good or bad stop could be determined by any number of things, ranging from the driver doing a poor job on the way in, to the stop itself, or the launch away from the box.

Either way, the pit stop is a pressure point during the race. Having been fitted with a heart-rate monitor for one race, you could tell when the pit stops had occurred simply by looking at the peaks on my graph: they matched the spike at the start which, as I've said, is another moment in the race when the strategist feels a lack of control.

It's exciting because the pit-stop decision is the bit that you are largely responsible for – so, if you've made that call correctly, it's very satisfying. But the pit stop is also where it can go wrong – as will be evident to everyone as soon as your car leaves the pit lane, particularly if it is behind the car you really wanted to be ahead of. In those circumstances, you've probably waited one lap too long.

The worst scenario is created by a safety car suddenly appearing as soon as your car leaves the pits. The rest of the field will get a cheap pit stop, losing less time to the cars remaining on the track – with your car being one of them because he has just made his stop. Without fail, someone on the team will sigh, or hit the desk, or moan. Of the hundreds of decisions I've referred to, the pit-stop call is the most visible, and the strategist's reputation can take a hit because of it.

More often than not, there is no time to dwell on what's happened because the team's second car is coming in on the next lap. On occasions, the two cars could be in close company and the strategist must make the next instant decision about when to stop the second car and whether to stack them.

The release from the pit box is governed by approaching cars. I will monitor the pit lane and advise the race engineer if there is going to be traffic. He, in turn, will warn the driver. If the pit stop goes according to plan, then traffic is easy to predict. But if our pit stop is slow, or another team's stop is slow, life becomes more difficult. If the pit lane is wide enough for two cars, we will have discussed beforehand the possibility of releasing the car and having it run alongside another – in which case, the driver obviously needs to be advised against launching straight into the fast lane, as he would usually do!

If the car arrives with damage, the pit wall will do their best to see and comment on anything obvious on that side of the car, while the chief mechanic will do the same from the garage side. Before that, we will have done our best, both at the pit wall and in mission control, to monitor any video images of either the incident or the incoming car.

The driver could be coming in for a less obvious reason: someone else's visor tear-off having gone into the brake duct, say. (Drivers start each race with multiple transparent tear-offs on their helmet visors, allowing the tear-offs to be removed at intervals to keep vision clear. The tear-offs are released into the air stream over the driver's shoulder and can cause overheating on a following car if caught in the radiator or brake ducts.) A puncture could be another reason for an unscheduled pit stop. Any number of things can hamper your strategy. As mentioned previously, the car could be running excessively hot when a stop is due. In such a case, the options are to either do a slow pit stop on that lap, with the driver making a gentler low-rev departure from the pit box, or to wait a few laps for everything to cool down.

In the 2017 Azerbaijan Grand Prix, our drivers collided with each other. Both cars were heading for the pits to have the damaged assessed, one of them needing a new front wing. Then they swapped positions on the way in. This meant the tyres and front wing were being readied in the wrong order in the garage. As if we didn't have enough to contend with . . .

Wet races are invariably the most hectic, with decisions coming thick and fast on every lap. The process can be slower if the track is changing from wet to dry and you can see the dry line start to appear on the track. This allows a little more time to make a decision, although there will generally be an optimum moment to bring the car in for a change from intermediate to dry tyres. Choosing that optimum moment can be difficult. If another driver is the first to box and then sets fast laps, apart from being too late, you must react immediately because the threat of an undercut by a driver on dry tyres is extremely high.

Going from dry to wet happens much more quickly. Equally, there can be a sudden need to change from an intermediate to a full wet if the rain becomes heavier, sometimes within seconds. That decision always feels a lot more pressured because you're trying to read the weather radar while keeping an eye on lap times. It is on occasions like this that the pit wall can see the bigger picture, whereas the driver is embroiled in coping with what's around him. I was always fortunate that the drivers usually did what was asked without question.

The standout exception occurred near the end of the Russian Grand Prix in 2021, when the track went very quickly from dry to wet. You may recall that Lando Norris chose to stay out in the leading McLaren in the hope of reaching the finish rather than losing time, and possibly his lead, with a pit stop. The chance of Norris's first win went with him when he spun off. We told both our drivers to pit because we knew the start of the lap was wet and needed intermediates. Sebastian Vettel and Lance Stroll heard the call – but ignored it because the pit entry was dry. They thought we had got it wrong, only to realise a few corners later just why the call had been made. Too late, of course, to take advantage of being on the right tyre.

The tiniest bit of information can make a massive difference. When the first car to change tyres is on its out lap, you need to know immediately if conditions warranted the change. This is so critical that you can't afford to wait until they reach the end of the sector (when the timing screen shows the first timing split). You will search for what amounts to a GPS mini-sector, just to get a feel for their speed and how it stacks up. Is this tyre working, or is it not? If it is, our cars need to come in NOW. Acting on that first bit of data could save your drivers from doing another lap on the wrong tyres and throwing away two, sometimes three, minutes.

Make or break decisions can come in those massively changeable conditions. And if the ideal stop lap is missed at somewhere like Monaco, you are toast. The wet-to-dry decision can be pain or glory. It's as clear cut as that. For a strategist, these are really tricky races. Turkey in 2020 was one example. It didn't really rain during the race – but neither did the track dry completely. It was a case of spending the entire race watching every piece of every lap in miniscule detail, looking for a tell-tale change in lap time.

It is like driving on the road in the rain: you need full focus every second. That's how it feels on the pit wall; you can't let anything slide. The additional problem is that your models in such conditions can be so variable and inaccurate because we do not race in the wet that often. With track conditions and temperatures changing so often, your model from three laps ago is useless now.

The strategist needs to work closely with the tyre engineer, who will be watching tyre temperature to establish whether the tyre is too hot, indicating a drier tyre may work. Track temperature also needs to be taken into consideration when establishing if a particular tyre will cope: will it recover, or is it going to fade? You need to be 100 per cent aligned. If, say, you put on an inter to do the whole race, it is likely to need to run with a slightly lower pressure than if required to only do ten laps really quickly. If the driver is pushing for the whole lap, you want the correct pressure to be there at the start to allow the tyre to reach optium pressure and grip as soon as possible. But if you want to string it out for the whole race on a surface that may begin to dry and become hotter, that's a very different story.

There also can be occasions when you choose what turns out to be absolutely the right stop lap in theory but, in practice, the tyre change is negated by other circumstances. A good example of this occurred at Monaco in 2016, when we changed Nico Hülkenberg from a wet to an inter. We were 100 per cent on the right lap; Nico was immediately much faster. But this being Monaco (and the same can apply at any track), it is impossible to go off the drying racing line and onto the wet part of the track to overtake cars that are actually on the wrong tyre for the changing conditions. You need to be very careful because the dry line is ready for the inter, but the wet line is not. 'Hulk' got stuck behind a Williams, meaning the stop lap was wrong.

I found there was more pressure when making those decisions than any other. It's hectic and you know mistakes will be made. You want to benefit from that, rather than being the strategist who gets it wrong. I'll admit we had a few bad races in changing conditions but, generally, we made the right decisions – which was always a nice feeling, particularly if it was a very important call towards the end of the race.

Chapter 23

THE FINAL LAPS

Believe it or not, one of the most difficult parts of any race – good or bad – is holding focus in the final five or six laps. It's rare that anything happens at that stage; very few decisions need to be made. Unless there is an accident or a puncture, a pit stop is unlikely. There comes a point when, even if there is a safety car and the life of your tyres is limited, a pit stop would be out of the question because the race would not restart. After the sometimes frantic activity during the previous 90 minutes or so, this comparative period of calm allows the mind to wander.

Again, strange as it may seem, it is easier to maintain concentration in a race that is not finishing well because there is always the hope that something good might happen. Whereas if there is a potential win or podium finish in the offing, there is a high chance of being distracted by thinking about what could go wrong rather than focusing on the usual lap-by-lap process. Minds tend to drift, either through excitement or disappointment, but, whichever end of the scale, the strategist tries to keep everyone together and thinking about the job in hand.

The drivers don't help. If running on their own with less to think about going into the final laps, they start imagining things. You'll get questions such as, 'Are the tyres okay?' Or, 'I think I heard something strange. Is everything okay?' To give peace of mind, the strategist needs to repeat these notions out loud and go through a check process by directing questions you wouldn't normally ask. And I must admit that the comparative struggle to get through the final laps is not helped by the sudden realisation you need to go to the bathroom – particularly if the race has been extended by lengthy safety car periods or, perhaps, a red flag!

And then there is the final lap. You are thinking about stuff that might happen, or incidents that could catch you out. An interesting situation sometimes arises where you are getting tight on fuel. You would be working out how much the pace could be reduced without losing positions to people behind. Or, if you are the last car on the lead lap, thinking about the possibility of dropping behind the leader and doing one lap less. Getting low on fuel has been rare in recent years, but there was a point in the past when we would push in the early stages, knowing fuel could be saved during the final lap.

A slightly awkward situation can arise if one car finishes quite some distance ahead of the other. One side of the garage is celebrating and doing all the necessary in-lap procedures (which I will cover shortly) while the other side still has half a lap, or whatever it might be, to go in the knowledge that this is not a done deal because anything could happen in the final few miles.

It is even worse if one car crosses the line moments before the winner takes the chequered flag. There are fireworks and all sorts of celebrations kicking off at the same time as your guy is continuing to race hard for one more lap. That could be difficult in somewhere like Mexico, particularly if, say, Checo (Sergio Pérez) was in for a top result in his home race. The grandstand opposite and the corporate boxes above the garage made so much noise, it was very difficult to hear what was being said on your headset, which could be crucial if your driver is in a close fight. That happened at Baku in 2016 when Checo was disputing third place with Kimi Räikkönen right to the end. Hülkenberg, his teammate at the time, was 42s behind, finishing the lap while the majority of the pit wall was celebrating.

My first podium with Force India in Russia came a year earlier, in 2015, at the end of a race with an eventful last lap. Bringing Checo in early under a safety car had seen us jump to third, a position he maintained for a long time. Towards the end, Checo held on as best he could before being overtaken by Räikkönen's Ferrari and the Williams of Valtteri Bottas, both of which were on fresher tyres. On the last lap, however, Bottas and Räikkönen collided and handed the podium place back to Checo. There was obviously a lot happening to hold your attention, unlike our win in Bahrain (which we will get into at a later stage in the book) when we were about ten seconds ahead of the pack. Then it was all about trying to sit on your hands, telling yourself it was going to be okay and there was no need to get involved.

Close racing throughout the field can vary from year to year. With Force India in 2020, we had the fourth fastest car, with a margin large enough to those behind to make the race feel lonely at times. When doing your models, it was a given that Haas or Williams would be slow enough at the back to be outside your pit-stop window very quickly and do away with the risk of getting caught behind them when rejoining after a stop. But for many years, the midfield has been very tight: the slightest error certain to cost places. If you made a mistake on the pit-stop lap, you wouldn't have the pace to correct it. You were fighting more competitors within a close range of pace, which meant there were no longer cars that would be comparatively easy to overtake.

That has translated into more split-second finishes down the field in recent years. Once your drivers have taken the chequered flag, the strategist will make a quick check to ensure there have been no penalties handed down for anything from an unsafe pit-stop release to a driver being blamed for an incident. The race engineers will be giving their drivers a run through the finishing order. This is the point when the strategist must be ready for questions from the cockpit about how such-and-such happened, or why, in the driver's opinion, a particular strategy was employed to his disadvantage.

Meanwhile, there's a lot to do on the in lap. Apart from a check on engine temperatures, the drivers will be reminded to pick up rubber in the form of the tiny marbles that have come from tyres during the race and are fringing the racing line. The reason for the driver going off line to allow his hot tyres to pick up this debris is because the car is the lightest it has been all weekend. Everything from fuel to brake material, coolant, driver fluids and tyre rubber – basically anything consumable on the car – will be at its lowest level. Picking up rubber debris will add to the car's weight and ensure it has not dropped below the permitted minimum. Normally, that should be unnecessary but, with a contravention of the weight limit bringing disqualification, it pays to be sure. While taking this precaution, drivers will be going slowly because, if your calculations have been correct, there should not be enough fuel left in the tank for a quick in lap.

Almost immediately, the analysis will start. 'Do you think we got that pit stop right?' 'What would have happened if we had done X, Y or Z?' The strategist will be doing some quick calculations to provide answers. It's never a great analysis at this point, but it does allow you to say something and, if necessary, promise to investigate the questions in more detail.

At this point on the pit wall, regardless of the finishing

positions, there would be a handshake, a thanks, a 'well done'. That is obviously difficult when you have had a bad race, but it is important when trying to maintain morale. The same applies when moving through the garage. Some of the crew members will be busy with breaking down the pit-stop equipment and the garage itself, but I would try to go to as many as I could and say thanks. At the same time, someone from hospitality will be handing out ice creams – always a nice touch, particularly after a hot race.

The post-race debrief is obviously the next big item on the agenda but, if both drivers have finished the race, it's necessary to stand around the coffee machine and wait for them to return from the media interview pen. Some people will be having lunch, because they didn't find time before the race. Others will be chatting about the race: 'Did you see the replay of that incident at the start?' Or, 'Did you hear what so-and-so said?' Meanwhile, I and others will have taken the welcome opportunity to dash to the bathroom!

The overall mood in the garage obviously depends a great deal on the outcome of the race. Essentially, having both cars score points and either maintain or improve your championship position is what matters. That would always warrant a 'well done'. The higher the finishing position – say a fourth, fifth or sixth – the greater the celebration. At Force India, Racing Point and Aston Martin, finishing on the podium felt as good as winning the race; second or third meant as much to us as winning on a regular basis did to Mercedes, the dominant team. A podium would be so far ahead of our expectations that the mechanics would be climbing the pit-lane fence to see and greet our car as it took the chequered flag. At this point, I would be on the intercom, thanking everyone at mission control but barely being able to hear anything because of the hugging and shouting going on around me!

The garage celebration could also embrace a race which, although not a podium finish, had seen us move up a place in the championship, or one where decision-making and strategy had gone very well and produced an unexpected result. On occasions, those races would stand out as much, if not more, than a podium because of the incredible team effort.

On the other side of the coin, even after what had been a bad race, I would follow the lead of my manager, Tom McCullough, and thank everyone. In some ways, this would carry more weight than a victory celebration. A disappointing race would generally not have been the fault of either the garage or the pit wall; it was likely due to the car being off the pace or an incident beyond our control. The guys in the garage would have worked as hard as ever, and it was important to recognise that.

Such recognition was even more critical in the event of a bad pit stop, because you knew no one on the crew set out to make that happen. I would always try to speak to whoever was feeling upset and annoyed, reassure them and say we would go through the video of the stop, establish why it had happened and, if necessary, act on it. But that could be difficult because there might be a number of people chatting to that person, wanting to figure out what had happened. It was best to step back at that point and then find a moment for a quiet word of reassurance later.

For me, I always found it more difficult if the strategy had not gone well. My initial reaction would be to find somewhere quiet, sit down and figure it out; I wouldn't want to loiter in the garage then. It's very hard, when frustrated and disappointed, to put on a brave face.

Of course, while you are feeling a little dejected, the people in the surrounding garages could be in full-celebration mode. How you felt about this depended on which team it was. There was a period when Williams struggled to score their first championship point in a long time. When it eventually happened, you felt very happy for them because you appreciated how tough it had been, and just what this meant.

On the other hand, there was every chance your direct rival would be the team garaged next door because of your respective places in the previous year's championship. In the event of them having a good result on a day when you had struggled, it would be a case of, 'Aw, shit. We could do without this right now!'

There could also be contradictory feelings at times. Esteban Ocon had raced with Force India/Racing Point for three seasons before moving to Renault and then, in 2021,

Alpine. In Hungary that year, Ocon scored his first win; a huge moment for any driver. Because we knew Esteban so well, we were very happy for him. I remember him looking at the group of Aston Martin people, because he probably felt more aligned to us than his new team; no disrespect whatsoever to Alpine, but that was a simple fact. So when celebrations are kicking off next door, your feelings depend very much on who is celebrating!

There can be occasions when, although not having achieved much as a team, you can be celebrating for someone else – but surreptitiously. In 2017 we were in a close fight with Williams for fourth in the championship; every point was vital either way. Towards the end of the Azerbaijan Grand Prix at Baku, the Mercedes of Valtteri Bottas beat Lance Stroll's Williams into second place by fractions of a second. It had been a close fight and it meant three points less for Williams – and a better result for Force India. We were doing everything to try to look calm, making sure our muted celebrations were not picked up by the TV cameras!

The post-race debrief is by far the longest of the weekend. There is a lot of data to go through, and there's no rush to get ready for the next practice session or to crack on with the analysis. You will go through a lot of the main strategy decisions, with the emphasis on the reasoning behind them. This is very important for the drivers because it will be their first insight into why we stopped on a particular lap, or precisely what triggered a change of plan. The drivers, in turn, could come back with how things looked from their point of view. It was always an interesting conversation for all sides, and everyone was totally committed to it.

The post-race debrief is also the first opportunity for the strategist to say – and this is very important – what, given the information we now have, you would do differently if you could run the race again. You outline targets for analysis: topics such as what would have happened if, say, we had stopped twice instead of once. The meeting gets people thinking and formulating their own ideas. Each department – particularly the drivers – speaks in more detail than they have at any other point during the weekend.

The race has provided the biggest bulk of long-run data for the car. Added to which, a rough estimation of everyone's fuel load at each point in the race gives very useful comparisons of pace. Having got such a good read from the race, this is your chance to improve things for the next grand prix. The inclusion of mission control in the debrief allows the factory to get a head start on issues while the race team is travelling home.

Once the debrief finishes, it's shut-down-your-laptop time and get ready to leave. You'll already have cleared your stuff from the pit wall because, by the time the debrief has finished, the pit-wall structure has gone and the garage is unrecognisable. You can't find your way around because the normal routes to and fro have disappeared under piles of boxes and packing cases. For European races, this is the easiest pack-up the strategist will do before heading for the airport.

Having established what went on in your own race, this is a good moment to take a quick look at how the results of others have affected your championship position. The fact is that you have been so engrossed in your own race, you often have no idea about how others have got on. That even extends to the team next door; if you haven't been racing them directly, you will not have a clue about the sort of race they've had.

The worst-case scenario here could occur when you arrived in the airport lounge on a Sunday night. This would often be the first time you had seen other team members face-to-face all weekend. Having previously worked for McLaren, I would say hello to the guys I knew very well and ask them how the weekend had gone, possibly completely unaware that they'd had a shocking race. 'Both cars retired by lap 5,' they would say. 'Did you not see what happened?' At which point, you immediately felt you should have been paying a bit more attention to the wider picture. I really disliked being in that lounge environment, trying to remember how everyone's race had gone. It would be difficult because, by the time you reached the airport, your head was so full of your own race and everything that had emerged during the debrief.

Among the many reports produced, the summary of the pit stops tended to be the most popular as the guys compared the times taken on each corner of the car. I would also work on the strategy report and a race trace showing how the race had developed. This would be important to factory staff, some of whom would arrive to work on Monday morning, not having seen the race or knowing anything about it. That may seem unusual to anyone engrossed in every detail as the race unfolded, but I recall one guy in the design department at McLaren. He had been there for years and I don't think he had ever watched a race. But he loved working in Formula 1 because this was design at the highest level. That may be an extreme case, but there would be a number of people involved with family or perhaps on holiday who wanted a snapshot of the race and a summary of where we stood in the championship.

The flyaway races are different when it comes to leaving the track. It will be necessary to ensure important reports and files are synced on your laptop to allow work during the long flight home. Instead of using a truck, you will have been working from a temporary office. There will be the need to pack up temporary tables, chairs and the accompanying bits and pieces, and there is time to do it because, usually, the return flights are not until the next day. In any case, it's good teamwork to help the guys who would otherwise be there for ages dismantling and boxing up all the garage and pit-stop kit.

It is surprising how quickly everyone gets into the routine of packing. Of course, there will always be someone who procrastinates and does less than everyone else, claiming, 'I just need to do one more email' – in which case someone just pulls the plug! The bottom line is that everyone wants to get the job done, pile into the bus and get back to the hotel. Depending on where you are in the world, you might then simply have dinner, or, in the case of somewhere like Austin or Montreal, head into town for the evening.

The prevailing mood when you get there depends on how the race has unfolded after many hours of preparation and anticipation. A fundamental attraction of Formula 1 is, of course, being unable to tell exactly how a grand prix will work out until such a highly charged and frequently emotional day is done.

Chapter 24

AFTER THE
RACE IS OVER

The Monday after a race – no matter how it had turned out – was never particularly good, mainly because I found it very difficult to do any work thanks to a mix of tiredness and coming down from a massive roller coaster of emotion, stress and energy.

There'd be a good conversation with the strategy team, who have been in mission control throughout the weekend, about tackling the various bits of analysis. I felt it was my responsibility (because the subject could be political) to do pace analysis and work out where our drivers sat in qualifying and the race. If one driver was being shown as a bit faster or slower than the other, it was important to be sure of your numbers! We would look at all the decisions made over the weekend, in a bid to establish how the perfect strategy model looked now that the race was over. How did the pace of the car compare with expectations? What did we get right? What did we get wrong? Was the tyre model as expected? Did the pit loss match pre-race estimates? The lessons learned would be the starting point for the next race and the following season.

Beyond the standard analysis, there are sometimes bespoke items relevant to that one race. It could involve going through video, or listening to intercom conversations, all of which have been recorded as a matter of routine. Wet races tend to bring more confusion and misunderstanding than normal and lead to 'he said/she said' moments of blame. If necessary, we would transcribe the conversations and try to understand where the communication had broken down. Usually, strategy would do this because it was also quite political, but I always felt there was valuable learning to be had – for everyone.

Listening back on comms recordings from the 2021 Hungarian Grand Prix provided a real lesson for me. This was the race when a first-corner collision (involving, among others, Lance Stroll in the Aston Martin) brought a red flag. It had been raining before the start. As the cars returned to the pits, here was an opportunity to change Sebastian Vettel from intermediates to slicks because it looked like the track might dry. Seb remained on inters but I had been very strong – or so I thought – in my opinion that we should fit dries (which, as it turned out, would have been the correct decision). When I subsequently listened to the recording, I realised I had not been forceful enough in getting my point across. Here was a good example of having something firm in your own mind but not doing a good enough job making that clear to others.

It was a similar story during the dry/wet conditions at the 2021 Russian Grand Prix (referred to in a previous chapter). Having told Sebastian and Lance to box for intermediate tyres, they both ignored the call because the pit entry was dry and they didn't see the need to change to inters. Listening to the recording brought the realisation that we had not placed enough emphasis on the pit entry still being dry while the rest of the track was rapidly becoming wet.

I didn't restrict myself to reviewing our race-team communications. I would always find time, post-race, to listen to discussions between the pit wall and the other 18 drivers. This flow of information is public and I found it to be one of the most interesting parts of my job. It's not as onerous as you might think because, when the blanks are removed, each conversation collapses to about 15 minutes. It was easy to do this on long-haul flights home: you could make a few notes around each team, and no data was necessary. You learn a massive amount about whether a team planned a one-stop and converted to a two, why they stopped when they did, which rival they seemed particularly interested in, how they reacted, and how happy – or otherwise – their driver was in the car. I would always listen to Kimi Räikkönen first because his clipped, direct comments cheered me up at the start of the day!

Checking out rivals' communications also provided an

insight into how they might react to certain situations in future races. It became apparent a few years ago that Alpine were listening very carefully and making very, very late pit-stop calls based on our radio communication. We became aware that, if racing directly against Alpine, it was advisable not to discuss pit-stop plans with our driver until the last minute, otherwise Alpine would react and possibly negate our strategy. Although it soon became common knowledge, we were aware of Ferrari's hesitant responses to certain situations, and we also learned that some teams would not open a safety car window in the way that we would, making them slow to react.

All of this helped build an image of what was happening along the pit lane. Even if you felt everything had gone perfectly with your race, there were probably nine sets of mistakes to learn from the other teams. I felt this was really useful and made me better at what I did.

We would estimate the absolute maximum points we should have scored with the pace we had, while ignoring events out of our control, such as lucking into a safety car. We would also look at the championship points lost – and why. Was that because of strategy? Or was it a reliability problem, or something unexpected?

By working out what we thought each team could have done if running their best race, it would feed into the wider learning beyond what happened to our team. If a rival missed an undercut, or stopped too late, or had a bad pit stop, what could we learn from that? How could we avoid that issue or mistake ourselves? We would keep a score throughout the year of points lost through strategy and, at the end of the season, work out the championship position had these additional points been picked up here and there.

There would be pit-stop analysis for us and others, checking out the timing of each stop. Obviously, we would have the detail for our own pit stops, enabling a comparison of how the wheel change went on each corner and checking whether the driver stopped on his mark. Comparing this with the information gathered on other teams, it would be possible to see where you sat in an informal league table of pit stops that weekend and to check out if someone was having a bad or a good race. If there was a poor – or a good – pit stop elsewhere, a look at the video would offer useful information worth including in your report, the aim being to have consistent pit stops, not the fastest ones.

Throughout the year, we would be tracking car pace at every race. There would be trackers for everything, from pace and degradation to the compound choices at other teams; how many runs they did in qualifying and how the weather measured up against the prediction for that weekend.

Prioritising areas in need of improvement was particularly important during my early days at Force India. We didn't do all the analysis I've mentioned because there wasn't enough of us to undertake it all. But we tried to focus on the bits that were going to improve our knowledge going forwards. We weren't too worried about how fancy the report looked; it just needed to be done efficiently and openly. One of our strengths back then was the honesty associated with everything we did. We were never afraid to produce a report that said strategy lost the team five points, or whatever it might have been. It is very easy to generate a report that says everything was fine; that there was no way, for instance, we could have envisaged stopping one lap earlier. Adopt that attitude and you begin to weaken the learning curve and, ultimately, the trust of those around you if the response is always, 'Oh, my bit's all right.' That line of thinking means there's no drive to succeed, and it prompts others in the group to feel the same.

It is important to be good at saying, 'This is where I think we should have done better.' And then figuring out how to improve something, be it a revised piece of software or a different procedure. It could be, for instance, that two people were examining lap times in the wet – and nobody was checking the radar. In which case, this could be the moment to introduce a new procedure covering various roles in the wet. Or, if one car retires, then there is a need for a clear line of communication about what anyone associated with that car should be doing henceforth in respect of the car that is still running. Each time there is a fault, the priority is to avoid a repeat. But the starting point for that must be an admission that there was a failing in the first place.

There was no apportioning of blame at Force India – mainly because we didn't have enough people to finger-point. Over time, that ethos has proved to be very successful for the team. It keeps the trust within the engineering office and beyond. You hear some mad stories of people in other teams changing the preview report to match what actually happened. You don't learn anything from that approach.

Following the immediate post-race debrief, there would be a much bigger one at the factory, probably on the Wednesday or Thursday. Whereas the debrief after the race would identify areas to look at, the follow-up in the factory would look to produce as many solutions as possible. Meetings with other groups could highlight the need to make certain changes, such as modifying the software because, perhaps, it was difficult to understand, or there was a specific issue. There could have been errors or problems with some of the tools, in which case you will apply as much correction as possible. Or it might be necessary to check the regulations and try to figure out why a penalty was handed out for no apparent reason.

You can find a strange mix of emotions back at the factory, depending on race results and what everyone's been through. There can be a particular disconnect when the race team returns from a double- or a triple-header, not having been with their families for a number of weeks. You're trying to slot back into factory life, catching up with people after not having seen them for some time – and yet you have all been working towards this common goal each race weekend.

The post-race procedure at the factory can vary from team to team. In the event of a podium at McLaren, everyone gathered for a glass of champagne and a chat about what had happened during the weekend. For a long time at Force India, the factory would get together for a round-up following a podium finish. Occasionally, regardless of the result, everyone would meet for a 'state of the nation' address, usually delivered by the team principal (sometimes after he, in turn, had been briefed by strategy). This was useful because a number of factory personnel, operating in their own little bubbles, might not be on top of a fundamental issue with the car, or they might need to know about global team news, the latest development path or the effect of a change to the regulations.

As I said, at no stage would there be any evidence of blame culture. It would be a case of, 'This is what we think went right – and this is where we went wrong.' This worked both ways, because neither would there be any 'credit to so-and-so for doing an amazing job'. Everyone would know where we stood as a team and could then crack on without any sense of jealousy, bitterness or recrimination.

As time went on, I found that the email list for the post-race strategy report got longer and longer; nobody asked to be removed from it which, in a way, was rewarding. Occasionally, someone would stop by my desk and ask about an item they didn't quite understand. After almost every race, there would be a little question you weren't expecting. I appreciated this because I could get further feedback from that. It also showed people were paying attention!

Every department produced their own analysis and, at some point, strategy would be asked to feed into that. The tyre engineers, for example, might have a question about temperature at a certain point in the race; in which case, thanks to strategy having better software for this sort of thing, I would be able to look at traffic on those laps and hopefully provide a useful pointer.

The final strategy report to wrap up the race weekend would be produced, hopefully before everyone on the race team took Friday as a lieu day at the start of a long weekend. It would be very different, of course, in the event of a double-header: the final report might not contain everything we would ideally want, or it might have items in need of further review.

Whatever the case, the driving force through everything above would be to improve, learn and do better next time. When the opportunity arose, you needed to be in the best possible position to win. All of which would make the moment so fantastic when it finally arrived.

Chapter 25

MY ONLY WIN

There was a point, early in 2020, when I didn't think we would have a grand prix season, never mind win a race near the end of it. But that's what happened as Sergio 'Checo' Pérez and Racing Point produced a truly memorable victory in the Sakhir Grand Prix on 6 December.

The first grand prix of 2020 had been scheduled for Melbourne on 15 March. When we reached Australia, there was talk of Covid-19 but, in our minds, the outbreak was restricted to China. At worst, it would mean the cancellation of the Chinese Grand Prix. By Friday morning, however, the potential seriousness of Covid was becoming apparent as someone at McLaren tested positive and the team immediately decided to go back to the UK.

Rumours became fact when the race weekend was cancelled and we were told to pack up and go home. But it shows how far away we were from realising the seriousness of the pandemic when, never having had the luxury of a Friday night off in the city of Melbourne, everyone went out together. We flew home on the Saturday by some convoluted route thanks to the last-minute travel arrangements. This was to be the start of a very tough and different year.

Initially we were furloughed at home. This may not have been to everyone's liking, but I really enjoyed having a bit of downtime not normally associated with that time of year. Every effort was being made to get a grand prix season going, the problem being finding venues and working out how it could be done safely. Back-to-back races at the same circuit were one answer, the first two being in Austria at the beginning of July. This, in fact, was also the first triple-header because we moved on to Hungary a week later, followed by another triple-header in August at Silverstone (two races), and then Spain. F1 was up and running, with 17 races being packed into what was effectively the second half of the year.

On the practical side, 2020 also became the year of the Covid test. Along with everyone else, I found that tough: PCR tests needed to be done at some point between a day and three days before travelling to each race. It meant driving to the factory for the test on what was supposed to be your weekend off. On top of which, we had to test every two or three days when at a grand prix, all of which made the crowded schedule seem even more intense.

So I remember 2020 as a tough year but, looking back on it now, Racing Point actually had a pretty good season. The results show that Checo's worst finishing position was tenth, with Lance Stroll regularly scoring points too.

Perhaps there was disappointment among some in the team who were frustrated by having a very good car but being unable to race it as often as we would have liked in the restricted calendar. But the fact was we scored in every race bar one in a season when we generally had two fast cars (Mercedes and Red Bull) ahead of us and then (except for McLaren) quite a gap to the rest of the field (Ferrari having made a strong start before falling away quite quickly). Having that margin allowed strategy decisions that we might not otherwise have made.

We may have had a consistently strong car, but things had become difficult after the second race when Renault protested the Racing Point RP20 on the grounds that the brake ducts were too close a copy of those on the 2019 Mercedes. While I wasn't involved in the technical discussion, the subsequent uncertainty meant the threat of exclusion hung over us going into each race. From a strategy point of view, there was no option but to simply get on with the job and not be distracted by all the noise and media fuss.

Just before the fifth round (the second race at Silverstone) in August, Racing Point was hit by a 15-point deduction. That knocked us back a bit in the Constructors' Championship but, by the time the final triple-header came round in early November, we were back in third place, some distance behind Red Bull and Mercedes, but ahead of McLaren and the rest – proof that the RP20 was probably the best car we've ever had. Now to see what we could do with it in Bahrain.

We obviously knew Bahrain very well but this visit for back-to-back races promised to be very interesting because we would be using two different tracks. The Bahrain Grand Prix on 29 November would be run on the customary 3.36 mile Bahrain International Circuit, while the Sakhir Grand Prix a week later would use what is known as the Outer Track. At 2.2 miles, the obviously much shorter circuit does away with the loop in the middle. (I'll refer to this as the Sakhir circuit, to avoid confusion.) The two tracks may be very different – but everything else is exactly the same from the team's point of view. Your 'home' hasn't changed; the pit wall and garage are the same; the start/finish straight and the first few corners are as before; you are continuing to stay in the same hotel. The danger here is that everything being so familiar subconsciously prevents the mindset adjustment you need when dealing with what is effectively a new track – and one we had not raced on before.

Under normal circumstances, moving from one track to another with dissimilar demands would be taken into account during all the pre-race briefings. The different strategies would be emphasised if, say, we were going from Monaco to Austria. For the former, track position means more than having a set pit-stop plan (ensuring you stop ahead of whoever is behind to prevent an undercut). Moving on to Austria a week later brings completely the opposite: choosing the optimum stop lap is everything. Even with the move from one country to another, it can be difficult to get everyone to adjust their thinking and accept that what was important last week is no longer relevant. In that respect, it is easy to see the potentially negative effect of a familiar workplace and routine, even though the demands of the two tracks could not be more diverse.

The Bahrain Grand Prix was a multiple-stop race. High tyre degradation meant it was easy to overtake as lap time bled away. The Sakhir circuit would have only 11 corners compared to 15 the previous week, leading to lower tyre energy which would mean less degradation, greater difficulty in overtaking and a one-stop race. This would be a fresh start; forget everything from the previous weekend.

It looked like we would want to forget the Bahrain Grand Prix in any case: Lance was eliminated on the second lap and Checo retired with engine failure. But the important point was that Pérez had been running in a strong third place (where he had been since the first lap) when the Mercedes power unit let go just four laps from the finish. Racing Point were looking good for the next race even though, as I've said, we were starting from scratch in many respects.

Recognising the different demands of the Sakhir circuit, we changed the downforce level on the RP20 – and were surprised to see that some teams had not made similar adjustments. We were coming towards the end of a 17-race calendar, having run 15 races in a very short space of time. Everyone in the paddock was tired, grumpy and, in many cases, one Covid test away from losing it completely. You could understand how easy it was to slip into a Bahrain mindset and get this job done without recognising race number 16 was going to be very different to number 15. For Racing Point, there was the extra motivation of payback for having lost a certain third place the week before.

The Bahrain Grand Prix had been won by Lewis Hamilton (his 11th of the season) but, the next day, he tested positive for Covid. George Russell would come across from Williams and take the reigning World Champion's place. Although Russell had done very little on-track running in the W11, he took a place on the front row, 0.026 seconds away from Valtteri Bottas in the other Mercedes. Checo qualified fifth, behind Max Verstappen's Red Bull and the Ferrari of Charles Leclerc on the second row. Lance qualified tenth.

The rules said the top ten had to start the race with the tyre they qualified on – which meant all ten would be on the soft for the first stint. With such a short lap in terms of time (pole was a very quick 53 seconds), the optimum strategy was a one-stop race, running the medium, followed by the hard. The challenge would be coaxing the soft tyre to survive a long first stint in order to make the preferred one-stop strategy work. For teams remaining in the Bahrain Grand Prix mindset, two stops seemed the most likely option, even though it would be the slowest way to run the 87 laps. The point was, the Sakhir track had no history to fall back on; no clues as to what might happen in the next hour and a half. As the 20 cars came to the grid for the start in the twilight, we were in for an intriguing race.

Sakhir Grand Prix, Bahrain, 6 December 2020

Turn 4 of the first lap, and Checo is involved in a collision that takes out Verstappen and Leclerc. I can see from the onboard picture that our car is at a standstill and facing the wrong way. Two things happen: a safety car is called; and Checo is quickly on the move. Forget about the incident; we need to know what damage, if any, has been caused to our car. With such a very short lap, there is less than a minute to decide what to do next. Is the car able to continue? Or do we need to bring him in and change tyres?

We've got a system in place with various people looking at lots of different aspects of the car. The aerodynamicists are checking the wings and the floor and looking for any drop in the pressures measured by onboard sensors. The tyre engineers are monitoring the health of the four soft tyres. Someone is watching the video rerun to assess collision damage – if any. And, of course, there is feedback from Checo himself. The safety car, by slowing the field, has bought us a fraction more time.

Word comes through that everything on the car looks good; Checo can keep going. Now it is a simple alterative: do we stop him, or not? My immediate thoughts from a strategy point of view are: we know that the fastest race is not the soft tyre we are on; Checo is running dead last, meaning little will be lost if we bring him in. The pit crew will be ready with the medium – the bail-out tyre (as discussed in a previous chapter) – which means the job can be done very quickly and allow Checo to catch up with the rest of the field, which will be running behind the safety car. I decide to stop even though, strictly speaking, I don't need to. It is a big decision. And yet it is one of the easiest decisions I've ever had to make. I'm absolutely sure the medium will be better than the soft we are on. It will give us the best chance of targeting the favoured one-stop strategy. When Chris Cronin, Sergio's race engineer, looks at me and asks, 'Are you sure?' I have no hesitation in replying, 'One hundred per cent.' It is the first step in putting us in a position to win this race.

There are other outside factors affecting how this could play out. Checo is keen to make up for having missed out on P3 last week. But, more than that, he has been told that he is going to be replaced by Sebastian Vettel next season. As things stand, Sergio Pérez does not have a drive for 2021. He has nothing to lose and a point to prove. After seven laps behind the safety car, the race is on once more. Checo moves up six places in four laps, latching onto the tail of Alex Albon and following the Red Bull when they overtake Vettel's Ferrari.

By lap 40 – almost half-distance – he is up to fifth, the Ferrari of Carlos Sainz and Daniel Ricciardo's Renault having already stopped due to being on a two-stop strategy which, we know, isn't going to work. Directly ahead are the Renault of Esteban Ocon and Lance, both of whom started on the soft. Ocon stops on lap 41 and we bring in Lance a lap later. Lance has done well to stretch the run on the soft tyre this far, but Checo is fit to go further on the medium. We extend his run to the end of lap 47 – which is a big difference in terms of what others are doing. Checo rejoins ninth, behind at least four cars we know are going to make a second stop because these teams, apparently not realising that low degradation would allow stopping just once, are locked into a slower two-stop strategy.

Checo, with better pace as the various strategies play out, overtakes Lance and, a lap later, Ocon. Now he is third. Russell leads Bottas, Mercedes being the only top team still running to have decided on a one-stop being the fastest strategy. This race belongs to them – barring misfortune.

With 26 laps remaining, Jack Aitken (who took over Russell's seat at Williams) spins at the exit of the final corner, touches the barrier and wipes off the car's nose. A virtual safety car, which freezes positions as the cars circulate slowly, becomes a full safety car when it is realised the removal of debris and gravel is more difficult than first thought and requires marshals on the track. This is an opportunity for pit stops for anyone on a two-stop strategy.

When they rejoin, they will have fresher tyres, but Checo's are not much older in terms of laps done and, crucially, he is ahead of them on the road.

But there's not much we can do about the Mercedes up front. Russell continues to lead from Bottas. Then it's Checo, Ocon and Lance. All five are on hard tyres and the correct one-stop strategy.

Before the safety car, Mercedes had a full safety car pit-stop window, meaning both cars could have stopped and remained in front. Now they see the safety car as an opportunity to fit a new set of tyres without losing track position. It is the right thing to do at this point. If they stay out and another safety car closes up the field with a few laps remaining, both Mercedes drivers will be at the mercy of those with fresher tyres, particularly Checo, given the speed and rhythm he's got.

So, the leaders dive into the pits. Their stops are a total disaster. Russell is fitted with Bottas's front tyres – which is not allowed. He comes back in 1 lap later for his allocated tyres, during which time Valtteri has had to wait while the pit crew fetch a full set of his tyres. The chaos has been caused by a glitch in the radio comms and the fact that the crew belonging to each car can listen to their own driver – which is fine under normal circumstances. But when a safety car suddenly appears, it's chaos. The drivers are talking; their engineers are talking; the strategist is talking. In this instance, the person on the pit wall making the call for the stops tells the whole crew that George is coming in. But at that very moment, Bottas is saying he is coming in – and his crew dealing with his front tyres do not hear the main message that Russell is about to stop. Valtteri's guys come out with his tyres – and the fronts go on George's car as he arrives first.

As the safety car withdraws, Checo is in the lead, with Ocon and Stroll still behind. Russell starts to come through, fighting his way into second place. But George then has to make another pit stop because of a puncture. Russell was close to having the speed to beat us – maybe not quite, but close. As for Checo, he has fresher tyres and better pace than both Ocon and Stroll.

There are nine laps remaining. With such a quick lap time, there are about ten minutes still to go to get to that win – ten minutes thinking about what might go wrong, wondering what the risks might be, checking everything in the car. Are we going to make it to the end? There should be no more pit stops. I can't do a lot in these final laps. I can be ready for another safety car, or should there be the need for a new front wing because of a collision. I'm thinking about what would need to be done in certain hypothetical situations. It could come down to either Checo making a mistake or to something breaking on the car. I immediately think of the engine failure in the previous race. Everyone's checking everything more diligently than they normally would, living in the hope that the next ten minutes are going to be okay. It feels like a lifetime.

Then come the intercom discussions: 'Tell him to slow down!' 'Why's he going so fast?' 'He's using too much kerb [and putting the car at risk].' 'Just make sure he knows he's ten seconds ahead!' I know where this concern is coming from thanks to the failure the previous week. But I also understand that Checo is pushing on because it's easier to stay focused that way. When a driver backs off, he loses rhythm and that's when mistakes are made. So, let him get on with his race.

In any case, the pit wall needs to stay focused because Lance is right with Ocon as they fight for second place. I'm trying to treat this like the closing stage of any race. But, in my head, I have the growing feeling that this is going to be special, very special indeed.

Into the final laps and I know that not even a safety car could wreck this now; there would not be enough laps left to restart the race. Going into the last lap, I'm not watching the data any more. All eyes are on the screens showing the onboards from both cars: Checo out on his own; Lance tucked under the rear wing of Ocon's Renault. There's even time for momentary annoyance that we somehow allowed Ocon to undercut us at his only pit stop.

But back to the leader. Our leader. Half a lap to go. There's growing excitement that we're nearly there. It's very difficult to remain calm as I become aware that the garage has emptied and everyone has run across to the pit wall. With eyes locked on the onboard screens, I can feel movement all around as the guys climb the fence and crowd around the sides of the pit-wall structure to get a view of the track.

As Checo appears on the pit straight and heads for the chequered flag, I hear the roar from the pit wall – and it continues as Lance sweeps past, one second behind Ocon. I'm bombarded by people – pats on the back, hugs, hand-shakes – the entire area is delirious.

Meanwhile, I'm checking the final positions – almost as confirmation that this has really happened. We've won by 10.5 seconds. That's a huge amount and simply adds to the truly fantastic feeling. Job done!

Many different things flash through my mind: working with all these great people as we got through a very tough year, living in bubbles, trying to keep contact to a minimum, dealing with a frequently strained atmosphere, particularly after losing 15 points. And now this massive achievement. It's my first win. We pulled together as a team to set up the car and plan the stop laps and strategy; everyone was aligned on this being a new race – a different race. We made it work. We won a GP!

Everyone rushed down the pit lane to get a good spot under the podium. The celebrations continued: hugging; shouting; raw emotion. It was crazy. Lance and Esteban came onto the podium. Then Checo appeared, joined by Andy Stevenson, sporting director of Aston Martin F1, to collect the trophy on behalf of the winning constructor. Andy had been with the team from the days of Jordan Grand Prix; a devoted and really hard worker, he was the perfect choice to represent the team as a whole.

It was a really great podium from our point of view. Apart from Andy and our two drivers, Esteban, having been with Racing Point for two seasons, allowed us to cheer as each trophy was handed out. Seeing this podium was particularly satisfying for me because Checo had not only been with Force India when I joined in 2015, but he had also raced with McLaren before that. We had done five and a half seasons together. Seeing him have his first win, and knowing I had played a part in it, was very satisfying, not to mention being something of a relief for Checo, given that he was out of a drive for 2021.

Since, in all honesty, our win was not expected, rival team members took the trouble to congratulate us as we made our way to the podium. As I walked away, once the winning ceremony was over, Ryan was waiting for me, excited and pleased that I had finally managed a win, knowing how much it meant to the whole team. That was really nice because, as a member of the Mercedes pit crew having just had a shocker of a race, being near such mad celebrations was probably the last thing he needed.

Ryan was not alone. Several of the management from Mercedes were in and around the podium waiting to congratulate us because, of course, Racing Point was a partner team and used Mercedes power units and components. Such generosity of spirit added to the warm feeling that comes with winning.

But no matter the result, business is business. The post-race debrief needed to be done. But after a race like that, it was very difficult to sit down and talk about track temperatures, tyre degradation, or whatever – particularly when you can hear the noise of celebrations in full swing outside the room. But the fact remained that the final race was taking place at the end of the following week in Abu Dhabi. We also knew Abu Dhabi would be Checo's last race for us and plans were made to give him the winner's trophy. This was quite a move because teams usually like to keep their trophies and, at the time of writing, the 2020 Sakhir Grand Prix has been the only win for the Silverstone team in its various guises since the days of Jordan Grand Prix.

Meanwhile, the celebrations continued into the Bahrain night. We went to a Mexican restaurant – lots of tequila, sombreros, and chaos. Because of Covid, team rules meant we weren't supposed to leave our hotel. But under what everyone agreed were exceptional circumstances, we went to Checo's hotel and had a party.

Toto Wolff, team principal of Mercedes, was there. He asked me if I would have made the final pit stop had I been in Mercedes's position. I said, from a strategist's point of view, stopping both cars did have an element of risk – but it was absolutely the right thing to do. I must admit, it was nice to be asked, and rounded off an extraordinary day. My first win; something quite a few who've been in this business for many years would love to achieve.

Chapter 26

MY LAST RACE

Sergio Pérez was replaced by Sebastian Vettel in 2021. I got on very well with Seb; he was extremely hard working and pushed very hard. Racing Point F1 Team became Aston Martin F1 Team and we suddenly had a lot more money so, technically, your job felt a lot more secure. But that also meant more pressure and expectation. Despite the continuing influence of Covid, there were 22 races that season. But it wasn't necessarily always fun; not like it had been before.

I found 2020 and 2021 very difficult. I reached a point in 2021 where I thought, 'There are a lot of races. Doing triple-headers means three weeks on the road and returning home so knackered that you're useless on the one weekend off. Something's got to change.' The relentless schedule also meant there was very little development of either yourself or the team; very little down time – apart from the few weeks during the winter once the racing had finished. Of course, you would reach the end of every year feeling tired, grumpy and ready to spend time at home. And then, by January or

February, you were ready to get back on the road again. It's the same cycle every year. But this was different.

During 2021 Pete Hall, senior strategy engineer, started going to some of the races. This meant I may not have been travelling but I was still doing my job; doing the hours, albeit from the factory. I often found it more difficult working in mission control because of having to drive home each night after a long day with no curfew and deal with details and glitches which just don't arise at the races where things such as catering are taken care of.

I got to the end of the 2021 season and assessed my situation. No one was dependent on me at home and I felt I could get by, even if I lost my job. I might not make enough money to continue enjoying my current lifestyle, but my attitude was that I would survive. Covering a race from mission control was not as easy as people seemed to think. I wanted weekends off to be able to go to weddings, birthdays and social events, which travelling, or a long race weekend spent at the factory in mission control, had made more or less impossible to attend.

During the 147 races since joining the team in 2015, I had not missed a single session. I began to realise that, particularly with the triple-headers, different time zones and so on, I'd rather be doing something else. Or, at least, be able to operate in a similar way to a growing number of mechanics allowed by their teams to choose a couple of races to miss in the increasingly crowded and busy seasons.

I wanted that sort of flexibility. By the time we reached the last race of 2021 I realised what I was feeling was more than the usual end-of-season tiredness and irritation. The following January, the inescapable feeling was that if I was unable to miss a couple of races, then I would go off and do something else. With no alternative work in mind, I felt I could save money during my notice period over the next six months and survive. Whatever the outcome, there had to be a break from such an intense existence at the races. I had been so busy that there had never been an opportunity to sit down and think it all through. Even if I had been able to consider it, there would have been no time to prepare a CV, apply for jobs or, say, enrol on a training course. The F1 job is so demanding that there is no time to pause and think about life.

Looking back on it now, deciding to leave was one of the best career choices I've made. Since I wasn't moving to another team and taking all I knew about Aston Martin with me, they didn't put me in a random job somewhere within the team. I worked my full six-month notice period continuing in the job I really enjoyed, as well as choosing the races I wanted to attend.

That six-month period was an eye opener. I found I enjoyed going to the track more. Little things that would have irritated me in the past no longer seemed to matter. I just got on with the job and made what seemed really easy decisions compared to before. Because it didn't have any repercussions on my career going forward, I found I was immediately able to say, 'Okay, let's do this.' But these were not off-the-cuff decisions; they were driven by the knowledge and experience that this would be the right choice for the team. Everything seemed totally logical and got to the point where Tom McCullough remarked that I sounded more confident than ever before when speaking on the intercom. There is something very powerful in not having to worry unduly. The imposter syndrome was gone and replaced with confidence that I was good at what I was doing and had control of my career.

My last race was the Hungarian Grand Prix on 31 July. That was an interesting experience, different in so many ways. Normally, I would be rushing from meeting to meeting, sorting out this, checking on that. In Formula 1 there always seems to be something that simply has to be done. There is no time to stop for a casual chat. But in Budapest I was able to reduce the pace and, for example, have a cup of tea with Will Hings from communications, who I'd travelled the world with, but rarely ever had time to hang out with. I did little things I had never found time for in the past, such as sitting on the top of the motorhome, taking it all in, thinking this could be my last weekend in the paddock, and appreciating that not everyone has the opportunity to be at the heart of F1 in this way.

Sunday was a very emotional day. People say lovely things – which they wouldn't normally say – when you're leaving. On his in lap after the race, Sebastian came on the radio and did a little shout-out, saying thanks and how nice it had been to work together. It really hit home that this was the last time I would be on the pit wall after 136 races in that seat.

Others in the team became quite emotional too. As I came away from the pit wall, I had a little cry. I'd stayed there longer than the others. I loved being a strategist and would have continued if I didn't have to face as many as 23 races in a year. This was the first time I had left a job without being forced to quit (as I felt had been the case at McLaren). It's a very poignant moment when you choose to leave a job that means so much.

Out of habit more than anything else, I did the debrief when, in truth, I really didn't need to because I would not be at the next race. One of the guys from catering made me a little cocktail, which was a first during a debrief! When we had finished, I left behind my laptop, phone and everything that was Aston Martin-related. With an empty backpack, except for a few personal items, I walked out of the paddock for the last time. I had contemplated staying on for Sunday night but decided that could become too messy. I said more goodbyes at the airport and headed for home.

I quickly realised that my experience in strategy could be useful elsewhere. One of the most interesting opportunities came about six months later with Sky Sports F1. The relationship had begun after I contributed to their *Any Driven*

Monday podcast, and I think they began to see the benefit of the race weekend being viewed through an engineer's eyes. Appearing on Sky Sports F1 was not something I had ever considered, but I enjoy learning about new things. Becoming a pundit on television would be very different – to say the least – but I felt it would be worth trying.

Rather than being focused on a single team, I found taking a more global view of strategy brought unexpected realisations. When working for a team, you work *across* the paddock. By that, I mean you go from hospitality in a direct line through the garage and onto the pit wall. You rarely see anything on either side. Working for the media brings a broader view as you work *along* the paddock, taking in all of the teams. Observing the F1 paddock as an outsider means seeing the whole story rather than being focused mainly on just two cars. Working with Sky Sports F1 is a very different team environment to the one I'd become accustomed to.

The media has its own demands and aims, and yet this had never occurred to me while busy working at the pit wall. How a TV programme is made simply didn't come into my thinking. It's been a fascinating learning curve dealing with terminology and systems that are alien to me; an eye-opener when working with creative people whose priorities are very different to those of engineers – and yet we're all in the same game.

Post-race conversations with home in the past obviously

gave me a feel for how much a viewer can pick up from coverage on TV. I could understand why some races, particularly those with multiple pit stops, can be hard to explain. That's why it's gratifying to be given the chance to offer a different branch of analysis within grand prix racing, as well as promote the world of engineering.

From the intense pressure of the strategy seat on the pit wall it was hard to imagine life on the outside looking in. It was a brave move to walk away. It has been very rewarding to see in a different light the sport I have worked my whole life in and enjoy so much. The next chapter is only just beginning.

Acknowledgements

Firstly, a huge thank you to Maurice Hamilton for the help bringing my random thoughts to life on the page. He also takes responsibility for the idea and concept of the book as well as the title.

Thanks to David Luxton and Associates for the guidance and encouragement throughout.

To Richard Milner and the entire Quercus Books team for the belief in the book and the support through the process.

To the many, many family and friends who have guided and supported throughout my life, education and career to date. In particular, my parents, Helen and Anthony and brother, Martin. There are too many others, past and present, to name but thank you all.

Many throughout my career have influenced or steered my direction to this point. All deserve thanks but special thanks to Prof Orr who mentored my Formula Student university project and pushed me to apply to McLaren Racing graduate scheme without which life would be very different. To Neil Houldey who was my mentor when I got there and to James Manning, transmission design team leader, for the support moving trackside. Mark Williams, Head of Vehicle Engineering at McLaren Racing for both organising the McLaren GT young engineer programme as well as the encouragement to join and beginning of life trackside. To the many friends made at Aston Martin Racing over the years that made life in F1 enjoyable; Oliver Knighton, Tim Wright, Chris Cronin, Daniel Priestman, Victoria Buxton, Will Hings and Jun Matsuzaki among some of the names. To Andy Stevenson and Otmar Szafnauer for their support throughout. Finally, to Tom McCullough for taking a risk on a non-strategy engineer and then providing the support and encouragement over many years to make it work.

To my new Sky Sports F1 family for accepting me into the group with open arms, especially Billy Mc Ginty, Director of F1, for taking the risk on a non-presenter!

Saving the best till last thank you to Ryan, my better half. Throughout my life in the paddock Ryan has been an ever-present source of solace and (often too honest) sounding board. Over the years he has often been the encouragement needed to travel the world outside the paddock and experience real life. Without his support, the courage and strength to step away from the pit wall would not have been possible. Let's see what lies ahead.

Index